AN INTRODUCTION TO COMPARATIVE BIOCHEMISTRY

AN INTRODUCTION TO COMPARATIVE BIOCHEMISTRY

BY

ERNEST BALDWIN, Sc.D., F.Inst.Biol.

Professor of Biochemistry at University College in the University of London, formerly Fellow of St John's College, Cambridge
Fellow of the New York Academy of Sciences

WITH A FOREWORD BY

SIR FREDERICK GOWLAND HOPKINS

CAMBRIDGE
AT THE UNIVERSITY PRESS
1970

Published by the Syndics of the Cambridge University Press
Bentley House, 200 Euston Road, London N.W.1
American Branch: 32 East 57th Street, New York, N.Y. 10022

Standard Book Number: 521 04095 7

First edition 1937
Reprinted 1937
Second edition 1940
Third edition 1948
Reprinted 1949
Fourth edition 1964
Reprinted 1966 1970

Translations available in Spanish, Japanese, Portuguese,
Dutch and Serbo-Croat

Printed in Great Britain
at the University Printing House, Cambridge
(Brooke Crutchley, University Printer)

For Pauline

Perhaps this Task, which I have propos'd to my self, will incur the Censure of many judicious Men, who may think it an over-hasty, and presumptuous Attempt....In answer to this, I can plead for my self, that what I am here to say.,...is premis'd *upon no other account, than as the noblest Buildings are first wont to be represented in a few* Shadows, *or Small* Models; *which are not intended to be equal to the chief Structure it self, but only to show in little, by what* Materials, *with what* Charge, *and by how* many Hands, *that is afterwards to be rais'd. Although, therefore, I come to the Performance of this Work, with much less* Deliberation, *and* Ability, *than the* Weightiness *of it requires; yet, I trust, that the* Greatness *of the* Design *it self, on which I am to speak,...will serve to make something for my* Excuse.

THO. SPRAT, *The History of the Royal Society*,
London, 1667

It is stranger that we are not able to inculcate into the minds of many men, the necessity of that distinction *of my Lord* Bacon's, *that there ought to be* Experiments *of* Light, *as well as of* Fruit. *It is their usual word*, What solid good will come from thence? *They are indeed to be commended for being so severe* Exactors *of* goodness. *And it were to be wish'd, that they would not only exercise this vigour, about* Experiments, *but on their own* lives, *and* actions: *that they would still question with themselves, in all that they do*; *what* solid good *will come from thence*? *But they are to know, that in so large, and so various an* Art *as this of* Experiments, *there are many degrees of usefulness*: *some may serve for real, and plain* benefit, *without much* delight: *some for* teaching *without apparent* profit: *some for* light *now, and for* use *hereafter*; *some only for* ornament, *and* curiosity. *If they will persist in contemning all* Experiments, *except those which bring with them immediate* gain, *and a present* harvest: *they may as well cavil at the Providence of God, that he has not made all the seasons of the year, to be times of* mowing, reaping, *and* vintage.

THO. SPRAT, *The History of the Royal Society*,
London, 1667

CONTENTS

FOREWORD

by SIR F. GOWLAND HOPKINS

THE publication of this book affords me much personal satisfaction, and to be associated with it on the easy terms involved in writing a brief foreword I count as a privilege.

That the student of biochemistry should, early in his course, be led to understand how important and enlightening are the facts to be won in the comparative field seems to me most desirable. History explains why this field has suffered relative neglect, especially in the teaching of the subject. Biochemistry as an independent discipline was an offshoot from classical physiology, and this, though for long the most experimental of the biological sciences, was mainly devoted to the service of Medicine, finding fields enough for its activities in the study of the vertebrate and particularly of the mammal. A generation ago inevitable departmentalism tended to separate the thought of the physiologist from that of the zoologist and botanist, whose pursuits from their very nature demanded the widest possible survey of animal and plant forms. Zoology and botany, however, though now becoming more and more experimental, were for long wholly descriptive sciences but little concerned with the active processes of life.

It is true that what we have come to call General Physiology is now a rapidly advancing branch of experimental enquiry, and it is perhaps less easy to justify an attempt to distinguish between its activities and those of modern biochemistry than between the latter and those of classical physiology. There must be, as chapters in this book make clear, an overlap in their fields which is entirely desirable. Yet there is still a distinction which seems to be real. Physiology as ordinarily understood is chiefly concerned in every case with the visible functioning of organs; biochemistry rather with the molecular events which are associated with these visible activities. I venture to think that productive thought in biochemistry in particular calls for the widest possible survey of life's manifestations. One of its ultimate tasks is to decide on what, from the chemical standpoint, is essential for these manifestations as distinct from what is secondary and specific. For any such decisions the necessary harvest of contributory facts must come from many diverse fields.

But biochemistry also aims at contributing its proper quota to the study of growth, evolution and heredity. It is striving to reveal the chemical differentiation which underlies morphological and functional differentiation, and it hopes to describe in its own terms the material basis of inheritance.

I think that those who are entering upon the study of a science should, before they are immersed in detail, be helped to survey its aims and its possibilities widely. There are always many paths to follow later on, and some, rather than others, may tempt them to persevere.

To this end, among others, my colleague's book contributes in admirable fashion. The literature of comparative biochemistry is highly diffuse, and scores of disconnected facts have been recorded in a great variety of journals, but Dr Baldwin has, I feel, selected material for his exposition with skill and judgment. He has made clear the aims of his branch of science and has dealt with its different aspects in nice proportion. He has illustrated very adequately in relatively few pages many of the main lines of current progress, leaving in the mind of the reader no doubts concerning their significance. Above all he has, I feel, displayed his own interest in the subject, an interest intense enough to be infectious.

I commend this book to undergraduates, and I also commend it to others whose interest may be great but their leisure small.

PREFACE TO THE FIRST EDITION

THE aim of this little book is twofold. In the first place it is hoped to provide an elementary text-book of this special subject suitable for students taking Biochemistry in Part I of the Natural Sciences Tripos, and in the second to provide a starting-point for those who, for any other reason, may find themselves attracted to the subject. In the former case, E. G. Holmes' *The Metabolism of Living Tissues* will already have introduced most of the biochemical concepts used in the present book.

I make no apology for the presence in this book of a good deal of matter of a morphological and physiological kind. Physiologists at any rate will not deny the desirability of maintaining a close contact between physiological and biochemical lines of thought. The cleavage between morphology and biochemistry appears to be deeper, but recent work on the chemical aspects of morphogenesis indicates that it is probably entirely artificial. Comparative biochemistry takes root in many domains of scientific knowledge and can only be approached from a very wide view-point. Its data lie scattered throughout almost the entire field of biological literature, and I have therefore given a list of useful books and reviews. A wealth of further references will be found in these, and the literature quoted in the present book is thus only to be regarded as a key to the general literature of the subject.

It is of course impossible to cover the whole range of the subject, even briefly, in a book of this size; all that can be hoped for is a discussion of certain selected topics. No attempt has been made to deal exhaustively with the higher vertebrates on the one hand or with unicellular organisms on the other. These, largely by reason of their clinical or economic importance, have been and remain the subject of numerous monographs and constant research. But the task of the biochemist is, after all, the study of the physicochemical processes associated with the manifestations of what we call life—not the life of some particular animal or group of animals, but life in its most general sense. From this point of view a starfish or an earthworm, neither of which has any clinical or economic importance *per se*, is as important as any other living organism and fully entitled to the same consideration, and unless such forms do receive considerably more attention than is accorded to them at present, biochemistry, as yet hardly out of its cradle, will assuredly develop into a monster. It is in danger of growing without properly differentiating, of developing members adult in size while possessing only the most primitive of supporting structures and correlating apparatus. Only by ensuring the proportionate growth of all its members can the full humanitarian value of biochemistry be realised, so that the policy of preferential endowment of research likely to have a more or less immediate clinical or industrial application is, at best, a very short-sighted one.

But fortunately there are other reasons for a more adequate consideration of these neglected animals. Although it is only now beginning to be possible to consider the reciprocal relationships between organism and environment from a physicochemical standpoint, the way has at least been opened. The physicochemical approach to evolutionary problems is peculiarly the business of comparative biochemistry, and probably few notions have had such wide repercussions upon the world of human thought as the theory of evolution. Interpretations differ, the records are fragmentary, yet they tell the most fascinating tale to be found in the literature of the whole world.

It is a great pleasure to record my thanks to Prof. Sir F. Gowland Hopkins for writing the foreword to this book and thus becoming godfather to yet another branch of biochemistry. To Dr Joseph Needham, whose enthusiasm went far towards awakening my own interest in comparative problems and whose advice has been invaluable, to Dr Sidnie Manton, who kindly read the manuscript, to Prof. H. Munro Fox, who read the proofs, and to my wife, who undertook most of the labour of preparing the book for the press, I must acknowledge my gratitude. Finally, my thanks are due to the Cambridge University Press and to Messrs Cassell and Company, Ltd., for permission to reproduce Figs. 6 and 9 respectively.

E. B.

November 1936

PREFACE TO THE FOURTH EDITION

DURING the fifteen years or so since the last edition of this book appeared there have been vast developments in biochemistry. There has been a less spectacular progress, but an increasing interest, in the comparative aspects of the subject. The still tender and delicate flowers of comparative biochemistry tend to be overshadowed by the massive and often rugged background of biochemistry in general so that to find and pluck them becomes a forbidding task indeed for the beginner. I think therefore that this little book may still have a place to fill and a duty to perform.

I am happy to record my thanks to Mrs Margaret Anderson, who prepared the index, and to Dr. D. C. Watts for many comments and suggestions, both of whom lightened the task of revision.

Like its earlier editions this is meant only to be an introduction to the field so that, except in matters of detail and necessary revision, I have not sought to change or expand its scope, but rather to preserve as well as I can the flavour of the first edition—a none too easy task after a quarter of a century.

E. B.

Northwood,
August 1962

NOTE TO CLASSIFICATION

It will frequently be necessary to refer to animals with greater precision than can be attained by words of everyday language such as 'worm', 'snail', 'shellfish' and so on. Students and others with no zoological training may find this troublesome at first, but it is hoped that the accompanying table of classification will prove helpful. It does not profess to be in any sense a complete system and in some respects might not receive unqualified approval as regards its accuracy. It has, however, the great advantage of simplicity and is published in this form in order to provide the student with a ready means of deciding to what phylum or class a given animal may be assigned.

CLASSIFICATION OF ANIMALS

PROTOZOA	Unicellular animals, e.g. *Amoeba.*
PORIFERA	Sponges.
COELENTERATA	Jellyfishes, corals, sea anemones.
PLATYHELMINTHES	Flatworms, many parasitic worms.
ANNELIDA (segmented worms)	*Polychaeta*, e.g. ragworm, lugworm. *Oligochaeta*, e.g. earthworm. *Hirudinea*—leeches. *Gephyrea*, e.g. *Sipunculus*, *Phascolosoma.*
ARTHROPODA	*Crustacea*—crabs, lobsters, shrimps, etc. *Insecta*—insects. *Arachnida*—spiders, mites.
MOLLUSCA	*Lamellibranchiata*—mussels, scallops, clams, etc. *Gastropoda*—snails, slugs, winkles, etc. *Cephalopoda*—squids, cuttlefishes, octopuses, etc.
ECHINODERMATA	*Crinoidea*—sea lilies. *Ophiuroidea*—brittle stars. *Asteroidea*—starfishes. *Holothuroidea*—sea cucumbers. *Echinoidea*—sea urchins.
CHORDATA	
PROTOCHORDATA	*Tunicata*—sea squirts. *Enteropneusta*, e.g. *Balanoglossus.* *Cephalochorda*—lancelets, e.g. *Amphioxus.*
VERTEBRATA	*Cyclostomata*—lampreys. *Pisces: Elasmobranchii*—sharks, dogfishes, rays, etc. *Teleostei*—most bony fishes. *Amphibia*—frogs, toads, newts. *Reptilia*—snakes, lizards, turtles. *Aves*—birds. *Mammalia*—mammals.

CHAPTER I

1. *Genesis*

NOBODY knows how, when and where life began. Moreover, although many authorities have attempted to define it, it has become progressively more and more difficult in recent years to know exactly what we mean by 'life'. No-one would deny that a dog or a tree is vastly more complicated than a stone or a nail; no-one would regard a crystal as alive or a bacterial cell as dead. But it has become necessary to think again about these things, and to think more clearly.

Bacteria are regarded as typically 'living' because they can grow and multiply at the expense of an appropriate nutrient medium. In the case of pathogenic bacteria this nutrient medium is provided by the blood or the tissues of the host they invade, and bacteria of this kind can carry disease from a sick to a healthy organism—a further demonstration of their 'livingness'.

Now diseases can similarly be transmitted by viruses; virus diseases can be conveyed from a sick to a healthy plant by inoculation with the infected sap. The amount of virus present increases as the disease runs its course and may reach 10 per cent or more of the dry weight of the tissues by the time the ailing plant dies. Because viruses can not only transmit disease from sick to healthy organisms but can also multiply, they were formerly regarded as comparable with bacteria, except in size, but unlike most bacteria they can only multiply in the tissues of their own specific hosts and appear to have no independent metabolism of their own. They are so small that they can pass freely through filters that retain ordinary bacteria and are still commonly referred to as 'filter-passing viruses' and considered to be *micro*-micro-organisms.

However, a new phase of interest began with the rather startling discovery that certain plant viruses can be isolated in crystalline form; they are, in fact, nucleo-proteins. Many of the viruses that attack plants, animals and microorganisms have now been found to be crystallisable nucleoproteins. This came as something of a shock, and not only to the bacteriologists and clinicians. Virus diseases can be transmitted by the crystalline virus material as well as by infected sap, and again the amount of recoverable virus increases as the disease runs its course. Here, therefore, we have something that has some of the properties of living material; like many bacteria, the viruses can transmit disease from one organism to another and, again like bacteria, they can multiply in the tissues of their hosts. Yet, at the same time and unlike bacteria, they seem to have no metabolism of their own and, unlike any other known kind of living stuff whatsoever, they are crystallisable.

These substances thus seem to bridge the gap between the living and the non-living worlds, a discovery that has had profound effects upon biological thought. It has been necessary, for example, to revise our ideas about the nature and the origin of life; indeed it seems that we can no longer use the word 'life' as a precise term because we now know less than ever exactly what we mean by it.

How can we now define 'life'? This has long been a difficult problem; now it has become more difficult than ever. As Pirie has written, '"Life" and "Living" are clearly words that the scientist has borrowed from the plain man. The loan has worked satisfactorily until comparatively recently....Now, however, systems are being discovered and studied which are neither obviously living nor obviously dead, and it is necessary to define these words or else give up using them and coin others.' With the discovery that viruses are crystallisable nucleoproteins it began to be necessary for the first time to realise that there is no fundamental gap between what is living and what is not. The effects on biological thinking have been profound and far-reaching. The Origin of Life, formerly no more than a pseudo-problem upon which time, thought, energy, ink and paper had formerly been generously and uselessly lavished, now became a true and a real problem, worthy at last of serious scientific consideration.

The Origin of Life has become something to be sought in times very far removed from our own; at some early time in terrestrial history. Already many stimulating essays on this important matter have been written, speculation has abounded, and a major Symposium on the subject was held in Moscow in 1957 with distinguished scientific contributors from all over the world.

To-day most geochemists believe that, in its early days, the earth's atmosphere was very different from what it is now. Nitrogen was probably present but there was little or no oxygen and not much water. Instead there were probably hydrogen, ammonia, methane, hydrogen sulphide, the inert gases and a few more; it was a strongly reducing atmosphere rather than the oxidising one we have at the present time. Furthermore, with no clouds and no ozone to keep it out, extra-terrestrial radiation of various kinds—electric discharges, X-rays, cosmic rays, ultraviolet radiation—must have been intense, and it has been pointed out by Urey among others that ultra-violet and other forms of radiation, acting upon this primitive atmosphere, must probably have led to the synthesis of large quantities of small-molecular organic compounds and the production of what has been called a 'primordial soup'.

Now Miller has shown—and his observations have been repeatedly confirmed—that among the products of irradiation of gaseous mixtures approximating in composition to that of the earth's primitive atmosphere, glycine and formic, acetic and succinic acids appear in high yields, together with a remarkably assorted collection of other organic materials. It can hardly be without significance that even to-day these are still the starting materials for the biosynthesis of many elaborate compounds—acetate for the synthesis of fatty acids, sterols and steroids, and glycine and succinate for that of porphyrins; compounds that occur to-day in living cells of every kind. But it is a far cry from this period of 'chemical evolution' to the biological evolution of living things and the gap between is hard to fill in except by speculation. It has, however, been said that, given enough monkeys, enough typewriters and enough time, one of the animals would eventually produce a typescript of all Shakespeare's sonnets. There seems to be no reason why this should not be true. Equally, given a large enough number of small molecules as letters of a biochemical alphabet and a few billion years to do it in, there seems to be little reason why random permutations and combinations should not eventually lead to the production of some primitive kind of organised system possessing potentialities for self-replication, survival and eventual evolution.

No doubt the monkeys would have produced some other interesting documents in the course of their efforts, documents corresponding to other kinds of organised systems; systems that failed to stand up to alterations in a constantly changing external environment, and which subsequently died out and left no trace.

Perhaps, then, there is no inherent improbability in the idea that, beginning with the simple products of irradiation of the earth's primitive atmosphere, larger and more complex molecules were produced by chemical evolution and that through countless random arrangements and rearrangements, accompanied by the formation of more and more complex substances, there eventually appeared a degree and a kind of organisation that carried with it potentialities for survival, reproduction and still further evolution.

Supposing then, that the period of chemical evolution eventually gave rise to some organised system possessing the requisite potentialities, organic evolution at last could begin. To our knowledge and understanding of this process biochemistry can contribute much, and hence arises the interest and importance of comparative biochemistry.

It is generally believed to-day that life, as we ordinarily understand the word, began in the sea. Many arguments have been advanced in favour of this belief, arguments too numerous and too diverse to be detailed here, but, as we shall see in the ensuing chapters, many of the facts about present-day life can best be explained on the supposition that this life was indeed cradled in the sea. From the first primitive living things there evolved more and more complex forms, some of which died out, some remained ocean-dwellers and others moved on to live in fresh water and even on the dry land. Changes in the nature of the environment have been associated with innumerable modifications in the patterns of life. Many adaptive changes of structure, behaviour and so on have been described, and it is the purpose of this book to draw attention to some of the biochemical modifications which have been associated with past environmental changes. And it may as well be pointed out here, at the very beginning of our argument, that we must not suppose that modifications of a biochemical or any other kind could be or ever were purposively accomplished. We must not think of an animal 'struggling' to adapt itself to a new environment, however hard it may struggle to keep itself alive. A given animal can or cannot become adapted to a new environment according as it does or does not possess certain quite definite potentialities. For example, the colonisation of fresh water by marine animals required, amongst other things, the animals' ability to maintain within themselves an environment in which life could continue and which would remain independent of the properties of the external medium. This involved the elaboration of mechanisms whereby the salt content of the blood could be kept at a level higher than that in the surrounding water. An animal incapable of evolving the necessary mechanisms could never penetrate into fresh water, and no amount of purposive struggling could make up for the lack of the requisite potentialities.

As we shall see, the colonisation of the fresh waters must have been an exceedingly difficult performance. Many adjustments and modifications were necessary before a marine animal could pass through the estuarine regions which separate the sea from the fresh water of the rivers. The colonisation of dry land too required numerous and far-reaching adaptational changes, yet to-day there are few places so barren that no living creatures whatever can be found in them. The insects have even produced an inhabitant for petroleum pools. And the migration of animals into new environments is still taking place to-day, new migrations being often assisted by wind, weather and the high speed of modern transport.

It is convenient to classify animals in terms of the places in which they live, and for this purpose we can divide them roughly into three groups, those inhabiting the sea, those which live in fresh water, and those whose habitat is the dry land. We thus get three major groups, the marine, fresh-water and terrestrial animals. But there exist certain very definite intermediate kinds of environments such, for example, as the littoral zone which separates dry land from the sea, and the estuarine zone separating the fresh water of the rivers from the salt water of the sea. To these might be added marsh and swamp zones such as separate fresh water from dry land in many places. It is a general tenet of evolutionary theory that changes in animals take place slowly and step by step, probably by a series of genetic mutations, rather than rapidly and all at once, and it is therefore probable that habitats of an intermediate type must have played a large part as barriers to the migrations of animals from one habitat to another. An animal which has lived under estuarine conditions, even for long periods, may develop the last of many prerequisite adaptations for fresh-water existence with almost dramatic suddenness.

We can best begin our study of these questions by comparing the properties of the three main habitats. The sea is so vast in bulk that its properties can change only very slowly. The specific heat of water is so high that large amounts of heat produce relatively small changes in its temperature, while its high viscosity ensures freedom from violent mechanical disturbances except where the water is shallow. On dry land, on the other hand, the daily changes of temperature are often of considerable magnitude, and on account of its slight viscosity the aerial atmosphere is liable to violent climatic disturbances. Water thus offers a habitat of great stability as compared to land, though there are many points in favour of the land as well. Air contains a much greater supply of oxygen than does water, while the possibilities of rapid movement on the land far exceed those in water because of the great viscosity of the latter. But with regard to stability and constancy of properties, water far transcends dry land. In this sense fresh waters are inferior to the oceans only in their lesser bulk, but there is another important difference here, this time of a physicochemical kind. Sea water contains roughly 100 times as much salts as does fresh water, and it is this difference in salinity that constitutes what is probably the greatest barrier to the colonisation of fresh water by marine organisms.

Now it is well known that the life of a particular cell or

tissue is very closely dependent upon the properties of the solution which surrounds it. If a rabbit's heart, say, is perfused with Ringer's solution it will continue to beat for many hours, but very small changes in the composition or pH of the perfusing fluid suffice to bring the heart to a standstill. The cells may still continue to 'live' for some time but their normal functional activity ceases. Now in the body of the animal itself, the properties of the perfusing fluid, in this case the blood, are very closely regulated. The total salinity, the relative proportions of the different ionic constituents, the pH, temperature and so on remain constant within very narrow limits throughout the life of the animal, and a large part of modern physiological zoology is concerned with studying the mechanisms whereby this constancy is maintained.

The most primitive living things can have possessed none of the complex regulatory mechanisms which we find in mammals to-day, though they may have been able to tolerate rather wider variations than can the highly organised, modern forms. Even so, the degree of constancy necessary for their continued existence could probably have been found in the sea and there alone. So long as animals remained in the sea there was no need for them to possess complicated regulatory systems; they could rely, for the most part at any rate, upon the constancy of their external environment for that of the blood and tissue fluids which make up the internal medium. But when they began to migrate into new environments, new means had to be found of maintaining an internal environment compatible with the continuation of their life and normal functional activity. It is our business to trace, so far as we can, the steps in the evolution of the self-buffering, automatic thermostat which is man. It is a colossal task, for, as has been said, 'we must interpret the past by the present'. We must deduce from the properties of modern animals the properties of those which died out in remote epochs and to which we owe our own ancestry.

The average man takes the constancy of his own temperature, the regulatory efficiency of his kidneys and so on so much for granted that he does not realise the difficulties of, let us say, the herring, which lives in a medium containing only some 0·5 volume of oxygen per 100 ml. as compared with the 21 vols. per cent present in air, and has an osmotic pressure about three times as great as that of its own blood, so that the fish must work constantly against the danger of dehydration. But had the fishes never succeeded in developing the relatively simple regulatory mechanisms which they do in fact possess, man might never have been possible; logical thought, speech and record might never have been evolved, and this book would certainly never have been written.

2. *The colonisation of fresh water*

It is an interesting fact that the number of species of animals living in fresh water is much smaller than that in the sea. Not a single representative of the Echinodermata, for example, is found in fresh water, and the Cephalopoda too are exclusively marine. Facts such as these suggest that only a selection of the many kinds of marine animals have succeeded in establishing themselves in the fresh waters. A number of reasons have been brought forward to account for this. One of the simplest is the fact that the temperature of fresh water is liable to much wider fluctuations than that of the sea, and this may well have had its effect in preventing the migration of animals particularly susceptible to thermal changes. But this is clearly only a very small part of the whole story. Another, and this time a more significant fact, is the circumstance that most marine animals hatch from their eggs in the form of larvae, often totally unlike the adult in appearance and, as a rule, very weak swimmers. Such larvae might well be swept away by the currents which are found in rivers. Moreover, in the sea, the larvae float about in the surface layers of the water and feed upon the microscopic green plants, the diatoms, which are abundant there. In fresh water on the other hand it is likely that they would sink on account of the lesser buoyancy of fresh water, and either starve or perish in some other way. But whatever the reasons may be, larval forms are of infrequent occurrence in fresh water, though there are very notable exceptions in the case of insects and Amphibia.

As a rule the young fresh-water animal remains inside the egg until after the larval stage has been passed, so that it hatches as a miniature edition of the adult and is capable of swimming against the prevailing currents, and, since the larvae of fresh-water animals are not self-supporting, the parent must lay eggs containing more food than is required in the case of a marine animal. This is beautifully shown in the case of a shrimp, *Palaemonetes varians*, of which there are two varieties. One, var. *microgenitor*, is marine, and the female lays about 320 eggs per annum with a mean diameter of 0·5 mm., while the fresh-water form, var. *macrogenitor*, lays only 25 eggs per annum, and these have a diameter of 1·5 mm. Although we have only a few measurements of the actual sizes of eggs, we know that as a rule fresh-water organisms lay fewer, and presumably bigger, eggs than do closely related marine forms. Thus the whelk, *Buccinum*, lays some 12,000 eggs each year, fresh-water gastropods producing only about 20 to 100. Similarly the oyster, *Ostrea*, spawns 1,800,000 and the fresh-water mussel, *Anodonta*, only about 18,000 eggs per annum.

Fresh-water animals succeeded for the most part in supplying the extra food necessary to allow larval stages to be compressed into the pre-hatching period, but this alone was not enough. Consider, for example, the case of the cephalopods. Although the young ones hatch in the adult form, indicating that the eggs must have been provided with enough food to last right through the period of development, no fresh-water cephalopods exist. Now it has been found that each egg of the cuttlefish, *Sepia*, contains only 0·8 mg. of ash at the beginning of development and no less than 3·3 mg. at the end, so that three-quarters of the total ash required must have been obtained from the sea. Developing sea-urchin eggs likewise obtain much of their salts from the sea water. Fresh water contains very little salt as compared with sea water, the latter containing at least 100 times as much as the former, and moreover, the dissolved solids of fresh water consist mainly of calcium bicarbonate (see Table 1), so that in order to secure a given amount of salts an embryo developing in fresh water would have to extract the salts from well over 100 times as much water as if it lived in the sea. In addition to providing a large supply of food materials, therefore, the would-be parent of fresh-water animals must equip its eggs with an adequate supply of salts. We can thus state the egg-requirements of fresh-water animals as: (i) Compression of larval forms, and (ii) Provision of adequate supplies of (*a*) food, and (*b*) salts.

Table 1. *Composition of some typical natural waters*

	Gm. per litre						
	Na	K	Ca	Mg	Cl	SO_4	CO_3
Sea water	10·7	0·39	0·42	1·31	19·3	2·69	0·073
Hard fresh water	0·021	0·016	0·065	0·014	0·041	0·025	0·119
Soft fresh	0·016	—	0·010	0·00053	0·019	0·007	0·012

When at later stages of evolution animals moved on to the land, yet another provision had to be made for the developing embryo, that of water. In egg-laying terrestrial vertebrates (e.g. birds, snakes and lizards) enough water to see the embryos through its development is put into the eggs when they are laid. The eggs of terrestrial invertebrates are usually very small and commonly get the water they need from the microclimates in which they are deposited, at the roots of plants, in hollow trees, under dead leaves and so on.

Some animals, even some which are still aquatic, did not succeed in making the necessary provisions for their offspring. Thus the Jack-shrimp, *Leander*, of the Norfolk Broads, which when adult is just as much at home in fresh as it is in salt water, has to go down to the sea every year in order to hatch its eggs, which will not develop properly in fresh water, which presumably cannot supply enough ash. Similarly the eel, *Anguilla*, ordinarily an inhabitant of fresh water, undertakes a migration of thousands of miles into the Sargasso Sea in order to spawn.

But the importance of the salt content of the natural waters does not end here. As Table 2 shows, the total salt content, and therefore the osmotic pressure, varies greatly from one natural body of water to another, and we shall return presently to a discussion of this factor. The relative amounts of the different salts also vary from one water to another as may be inferred from the data of Table 1, and a peculiar significance attaches to the composition of sea water in this connection.

Table 2. *Salt content of some natural waters*

	% Dissolved solids
Californian salt lakes	30
Sea water	*ca.* 3
Fresh waters:	
'Utah Lake', Utah	$1{\cdot}16 \,.\, 10^{-1}$
Hardest river water	*ca.* $3 \,.\, 10^{-2}$
Lake Michigan	$1{\cdot}18 \,.\, 10^{-2}$
Rain water	*ca.* $3 \,.\, 10^{-3}$

3. *The ionic composition of the blood*

If we compare the relative proportions of the different ions present in the bloods of a number of different animals, we find that there exists between them a very remarkable similarity. Had we taken random samples of different inorganic instead of biological materials, we should have been considerably surprised if there had been any particular resemblance between them, and if any marked similarity were noticed we should probably conclude that they were in some way related and perhaps even had a common origin. The data given in Table 3 show that the bloods of widely different animals are very like each other and closely similar to sea water.

This resemblance to sea water seems to have been noticed by Bunge about the middle of the last century but was first emphasised by Quinton in the early 1900's. Quinton believed that blood is nothing more or less than an elaborated form of dilute sea water. As we shall see, Quinton's ideas were too crude, but he was able

Table 3. *Relative ionic compositions of the bloods and tissue fluids of some different animals (after Macallum)*

	Na	K	Ca	Mg	Cl	SO_4
Sea water	100	3·61	3·91	12·1	181	20·9
King crab, *Limulus*	100	5·62	4·06	11·2	187	13·4
Jellyfish, *Aurelia*	100	5·18	4·13	11·4	186	13·2
Lobster, *Homarus*	100	3·73	4·85	1·72	171	6·7
Dogfish, *Acanthias*	100	4·61	2·71	2·46	166	—
Sand shark, *Carcharias*	100	5·75	2·98	2·76	169	—
Cod, *Gadus*	100	9·50	3·93	1·41	150	—
Pollack, *Pollachius*	100	4·33	3·10	1·46	138	—
Frog, *Rana*	100	—	3·17	0·79	136	—
Dog, *Canis*	100	6·62	2·8	0·76	139	—
Man, *Homo*	100	6·75	3·10	0·70	129	—

Note. These figures are those for plasma or serum, not for the whole blood. The blood cells are not taken into account since they cannot be regarded as a part of the 'internal environment'.

nevertheless to bleed dogs till they were nearly white and then make up the blood volume with diluted sea water without any very untoward consequences.

In Table 3 the amounts of the different ions are expressed as percentages of that of the commonest cation, that of sodium, the amount of the latter being taken as 100 per cent. In this way it is possible to compare the relative compositions of fluids which differ more or less in absolute composition. On account of their general resemblance to sea water, it was later suggested by Macallum, who had the advantage of much improved analytical methods, that the circulating fluids of all the animals originally came from the sea water of some millions of years ago; in other words, that our own blood is nothing more or less than a modified sea water. This does indeed seem reasonable if, as we believe, life really originated in the sea. But if this is so, why is it that, as Table 3 shows, our own blood serum contains relatively more potassium and less magnesium than sea water? Macallum answered this by pointing out that, according to the estimates given by contemporary geologists, the primitive or Archaean Ocean contained more potassium and less magnesium than is present in modern sea water. Since that time the composition of the ocean has been gradually changing on account of the precipitation of various compounds on to the bed of the sea and the washing down of others by river action, the nett result being a fall in potassium and a rise in magnesium. Macallum believed that the vertebrates originated in a so-called Eovertebrate Ocean, at a time when the composition of sea water was very like that of the sera of modern animals. Figures for the compositions of the Archaean, Eovertebrate and Modern Oceans are given in Table 4.

Table 4. *Composition of sea water at different periods*

Period	Na	K	Ca	Mg
Archaean (geologists' values)	100	100–250	10	0·01–0·1
Eovertebrate (Macallum's values)	100	6·7	3·1	0·7
Modern (after Dittmar)	100	3·61	3·91	12·1

If we look again at the figures of Table 3 we see that the body fluids of *Limulus* (king crab) and *Aurelia* (jellyfish) resemble modern sea water much more closely than do those of the other animals listed. Macallum interpreted this as an indication that the blood system of these animals became shut off from the sea more recently than that of the animals from which the vertebrates subsequently descended. His view was essentially this, that life began in the ocean many millions of years ago, that the blood systems of different groups of animals closed off at different times, and that the composition of the bloods of their descendants has remained practically unmodified ever since owing to the influence of heredity.

But ingenious though this hypothesis is, there is in it one very weak point. Macallum supposed that once the blood system had been shut off from the sea water, the composition of the blood remained the same through the operation of heredity. Now we know that if a marine invertebrate, such as the edible crab, *Cancer*, or the sea hare, *Aplysia*, is placed in diluted sea water, salts pass out of its body, while if an excess of, say, potassium ions is injected into it, they rapidly come out of the body into the surrounding water. Since the body surface is evidently permeable to ions, it is clear that the internal and external media could not statically remain different in composition for any appreciable length of time. Quite clearly some kind of active process would have to be involved, and Pantin suggested a dynamic mechanism by which the observed differences between the two media could be fairly easily maintained. He argued in the following manner.

Consider a system consisting of a salt solution, bounded by a membrane which is permeable to salts and water alike, and immersed in a very large volume of sea water. Since water and salts can pass through the membrane, it follows from the laws of diffusion and osmosis that the system will eventually come into osmotic equilibrium, both solutions having the composition of sea water. Now suppose that a pump of some kind is at work, continually removing some of the internal medium, for the purpose in the animal (of which this is a model) of carrying away waste products of metabolism. Fresh fluid will enter from outside to replace that removed by the pump and, provided that all the ions can pass equally readily through the membrane, the incoming solution will be pure sea water and the composition of the internal medium will remain unchanged. But suppose now that some ions move through the membrane more rapidly than others. The incoming fluid will then come to contain relatively more of the faster and fewer of the more slowly moving ions and the composition of the internal fluid will change. After the pump has been working for some time a new equilibrium will be established, the internal medium containing relatively more of the faster ions. The activity of the pump or excretory organ thus brings a new factor into play, namely the relative mobilities of the different ions, with the result that a new equilibrium, this time a dynamic and not a static one, is set up.

It has been argued that (*a*) more of the faster-moving ions could not enter, because to do so would mean moving against a concentration gradient, (*b*) that if slowly moving ions were left ouside, the internal osmotic pressure would fall so that the system would lose water and shrink accordingly. This, however, seems to be false argument. Osmotic pressure is determined by the concentration of dissolved particles in a given solution, irrespective of their nature. Osmotic equilibrium could be preserved as well by the entry of more fast-moving ions as by a loss of water.

Now it is certain that all the ions of sea water do *not* move at the same speed. The magnesium ion, for example, moves much more slowly than the ion of sodium; it carries with it a large shell of water molecules which considerably retards its passage through the medium. The potassium ion is less heavily hydrated than that of sodium and moves correspondingly faster. In the model, therefore, we should expect to find that the internal medium would come to contain more potassium and less magnesium ions in proportion to the number of sodium ions, and this is precisely what happens in the animals which the model represents—their bloods contain relatively more potassium and less magnesium than the sea water which constitutes their external environment. It is in fact true that, in general, living cells contain more potassium and less magnesium than is present in the liquid which bathes them, and in the extreme case of mammalian red blood cells, the internal medium contains large amounts of the fastest ion, potassium, and practically none of the other positive ions. The same is true of the cells of many other organisms, including those of the seaweed, *Valonia*.

Michaelis found that the potassium ion moves about 1·5 times as fast as that of sodium in aqueous solution, and this difference is evidently too small to explain the large differences observed between, let us say, the cell sap of *Valonia* and the sea water outside. But, generally speaking, biological membranes carry a negative charge, for the fluids bathing them are usually well on the alkaline side of the isoelectric pH of such membranes. Michaelis studied the migration of ions through charged membranes and found that the presence of charges of this kind has the effect of greatly exaggerating differences due to the different mobilities of the ions themselves. It is therefore probable that the presence of negative charges on biological membranes plays a large and important part in making possible the maintenance of concentration gradients between the media which they separate.

On this view therefore we may attribute the observed differences between the internal and external media to different ionic mobilities, exaggerated by the presence of negative charges on the boundary membranes, the whole effect being called into play by the activity of the excretory organ.

If this point of view is correct, we might expect to find that as the excretory organ becomes more and more efficient, the greater would become the difference between the blood serum and the waters of the sea, and this expectation is borne out in a general sort of way by the figures of Table 3. As we pass from simpler to more complex animals, the greater becomes the relative excess of potassium and the relative deficiency of magnesium.

The essential mistake in Macallum's argument was his supposition that, once established, the composition of the blood 'stays put' as it were of its own accord, whereas in actual fact it must be actively maintained by a dynamic process of some kind, and Macallum took no account of such processes. Moreover there is no doubt that an active absorption of ions can take place at the surface membranes of many fresh-water animals and similar activities appear to be an important factor in marine forms also. Such an absorption is certainly important during embryonic development (cf. p. 12).

Neither Macallum's theory nor that of Pantin can, however, explain the observed facts completely. Both points of view are certainly too simple to give a complete account, but they have at any rate brought to our notice a number of facts of the utmost importance, and by combining the two points of view and rearranging the argument a little we arrive at some very interesting conclusions.

The animals of to-day are the products of millions of years of evolution along widely divergent paths, but in spite of this their bloods are remarkably alike in ionic composition. This suggests that *the conditions under which cell life is possible are very restricted indeed and have not changed substantially since life first began.* It seems not improbable that the sea formerly contained more potassium and less magnesium than it does to-day, and that at one time its composition may well have been very like that of the sera of the animals of present times. Thus it seems probable that at some time in the past the composition of the sea was such that life was possible in it, and we may believe that fully organised life began at that time under the conditions then obtaining. *The subsequent evolution of different forms of life has necessarily been attended by that of mechanisms for maintaining within the organism an environment with the properties required for the continued life of its cells.* This essential truth is contained in Claude

Bernard's celebrated aphorism, 'La fixité du milieu intérieur est la condition de la vie libre'.

Thus, instead of being surprised that the bloods of different animals resemble each other so closely, we must realise that it could not have been otherwise. The composition of the blood has remained the same because the conditions under which life is possible have remained the same. It has remained so by no static process like the kind of fossilisation suggested by Macallum; it has been actively maintained, as indeed it must if life were not to become extinct.

We have just seen how in simple marine animals the necessary maintenance could be accomplished in its broad lines by means of an excretory organ of a fairly simple kind, advantage being taken of the fact that different ions happen to move at different speeds. We must realise, however, that, but for the fact that these ions happen to have the properties they do, maintenance of the necessary conditions might never have been possible. We can imagine that life of some kind may have been possible at other times and in other environments, and that living forms may actually have emerged, only to perish later on because the properties of the environment were such that, when external conditions changed, the organisms were unable to maintain the old conditions within themselves. Life not only requires that the living stuff itself shall have the right properties, i.e. be adapted to its environment—the properties of the environment too must be right, or life will sooner or later be extinguished; in other words, 'fitness' is essentially a reciprocal and not a one-sided relationship.

CHAPTER II

1. *The regulation of osmotic pressure*

THE maintenance of the proper 'milieu intérieur' was a relatively simple task for marine animals, for the concentrations of the internal and external media are not startlingly different even to-day. But as soon as animals began to move on from the sea towards fresh water, far-reaching changes in the regulatory mechanisms became necessary. Not only are the relative proportions of the different salts very different in fresh water from what they are in the ideal 'milieu intérieur', but the total salt content is very much lower. Animals were thus faced with the additional difficulty of keeping the osmotic pressure of their blood above that of the surrounding fresh water.

Animals such as crabs, lobsters, worms and the like can tolerate appreciable changes in the osmotic pressure of their blood, and this is presumably bound up with the fact that the osmotic pressure of sea water itself varies somewhat from place to place owing to evaporation and to dilution with fresh water, the extent of which differs from one region to another. But in more highly organised forms the limits of tolerance are much narrower. In the mammals, for instance, small variations in the osmotic pressure of the internal environment impair the efficiency of the cells, large changes destroy it completely. An interesting example of the importance of salt content is met with in miners, stokers and others who work for long periods at high temperatures. Such men sweat profusely and thus lose water and salts. If they drink plain water to make good their losses, they are liable to suffer from violent muscular cramps, but by using diluted brine or salted beer the cramps are avoided. The reason, apparently, is this. The chief protein of muscle, actomyosin, is insoluble in pure water but soluble in dilute salt solutions and reprecipitated by excess of salts, i.e. it is a globulin. The normal functional condition of this protein can only be maintained by keeping the salt content of the blood at a certain constant level; any disturbance impairs its efficiency, and if water alone is replaced after a heavy bout of sweating, the blood salts become slightly diluted, the physical state of the actomyosin is slightly modified, and cramp results. Clearly, therefore, the regulation of the total salt content of the blood is a very necessary part of the total regulation of the internal environment.

If a marine invertebrate such as the spider crab, *Maia*, is taken from the sea and put into diluted sea water, it is found that salts pass out of the body till the internal and external salinities are the same, i.e. till the osmotic pressure of the blood is equal to that of the water outside. Unlike a man or a rabbit, *Maia* cannot regulate the salt content of its blood, which is therefore always isotonic with the surrounding water. This crab normally lives in the sea which, as we have seen, has remarkably constant properties, and the animal relies on this external constancy for that of its own blood. Animals like *Maia* cannot withstand any considerable changes in external salinity and are said to be *stenohaline*. So long as animals continued to live in the sea there was no need for them to be anything other than stenohaline, but when, presumably by genetic mutations, their physiology happened to change in a suitable direction and they became able to tolerate larger external changes through some kind of control over their internal composition, they could begin to move shorewards towards estuarine conditions. Estuarine water varies from salt to fresh with every tide, and a way had to be found of withstanding wide fluctuations of external salinity before these estuarine regions could be successfully colonised. This accomplished, animals might pass on up the rivers and thence, eventually, to the land, provided no further obstacles barred the way.

The well-known shore crab, *Carcinus*, is a typical estuarine organism and would be expected to be capable of withstanding wide variations of external salinity, i.e. of being *euryhaline* as opposed to the stenohaline *Maia*. Two methods by which this could be accomplished suggest themselves; the tissues might be so modified as to be able to work at high or low osmotic pressures, or the animal might in some way maintain a constant internal salinity in spite of external changes. The first of these possibilities has been exploited by some animals, by the lugworm, *Arenicola*, for example, which flourishes even in the very dilute waters of the Baltic. But this method of adaptation to low external salinities seems never to have allowed marine animals to tolerate exposure to completely fresh water; the second alternative seems to have played the greater part in the conquest of estuarine and fresh-water environments.

If we take specimens of *Carcinus* and place them in diluted sea water, we find that the osmotic pressure of their blood falls a little, but far less than that of *Maia* would do under similar conditions. Even in practically pure water the salinity of the blood of *Carcinus* does not fall enough to kill the animal, so that this crab is able to live in the open sea or in estuaries, and even, on occasion, to penetrate a considerable distance up the rivers. A third possibility, the absorption of salts from the surrounding water against an existing osmotic gradient, is also an important factor in many cases. We shall return to this point later.

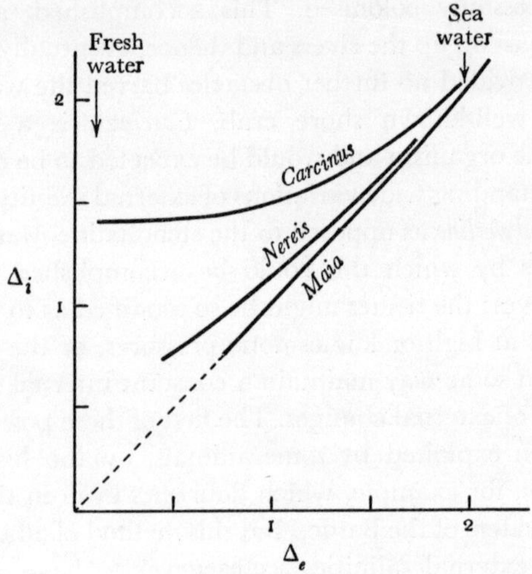

Fig. 1. Variation of internal with external osmotic pressure in some invertebrates.

The relationships between the internal and external osmotic pressures of *Maia* and *Carcinus* are expressed graphically in Fig. 1, in which the internal value, Δ_i, is plotted against the external, Δ_e. (Osmotic pressure is usually expressed in terms of the freezing-point depression Δ, to which it is proportional.) The straight line for *Maia* shows how, in the case of this crab, the internal and external salinities are always the same, the animal being unable to control its own osmotic pressure. The curved line for *Carcinus* on the other hand shows how this species maintains a high internal osmotic pressure even when that of the external fluid is very low, and is typical for euryhaline animals. There are, of course, different degrees of euryhalinity, and the third curve shows the behaviour of an annelid worm, *Nereis*, which is euryhaline, though less so than *Carcinus* and more so than the completely stenohaline *Maia*. The species in question is *N. diversicolor*. It has been found that specimens taken from regions in which the daily osmotic changes are fairly large are more euryhaline than those from regions where the external medium is more constant in composition. In this case it would seem that the wider fluctuations in the external medium call forth a greater measure of control over the internal medium—a racial rather than a specific difference. Another species, *N. cultrifera*, is almost completely stenohaline.

Before an animal could pass through the estuarine regions on its way to the fresh water of the rivers, it would need to become euryhaline. It must be able to maintain a constant internal salinity, no matter how the external may vary. Having reached fresh water, however, the ability to withstand considerable changes of external osmotic pressure would no longer be necessary, and the range of external variations to which the animal could adjust itself would therefore be expected to diminish, until the maintenance of the necessary internal salinity was only possible so long as the animal was in its natural environment, fresh water. This is what has actually happened. Fresh-water organisms, though osmotically independent of their normal environment, are nevertheless stenohaline as a rule—their regulatory powers break down if the osmotic pressure of the external environment increases appreciably. The osmotic independence of fresh-water organisms is all that remains of the euryhalinity which their ancestors must have possessed when first they penetrated into the fresh water. The majority of animals, fresh-water and marine species alike, are thus stenohaline; the only organisms which are euryhaline to-day are migratory forms such as the eel and the salmon, and estuarine types such as *Carcinus*.

Let us now consider in what manner osmotic independence can be attained. The problem is an interesting one if only because, had the early animals not succeeded in finding its solution, fresh waters and the land would be at any rate much poorer than they are in living forms, and man would probably never have come into being.

Take first of all the case of a stenohaline marine animal such as *Maia*. Here the surface membranes are permeable to water and salts. If this animal is exposed to sea water more concentrated or more dilute than that in which it normally lives, or if an excess of salts is injected into it, water and salts pass through the membranes till osmotic equilibrium is reached between the solutions on either side. Now the loss of salts to a weaker external medium could be prevented by making the membranes semi-permeable, i.e. permeable to water but not to salts. But this would be jumping out of the frying pan into the fire, because, as we know from the laws of osmosis, water passes from the weaker to the stronger of two solutions separated by a membrane of this kind, and water would therefore enter the body of the animal. The animal would have to swell in order to accommodate the incoming water, and the process of swelling could not go on indefinitely without bursting the animal. This in its turn might be avoided if the membranes could be made impermeable to water as well as to salts, but probably this is not a practicable proposition, since the animal must be able to obtain oxygen from the surrounding water and to discharge into it carbon dioxide and other waste products of its metabolism. The necessary degree of permeability to gases appears to have been inseparable from an appreciable permeability to water, and it seems that animals which are osmotically independent do indeed have semi-permeable membranes and the danger of bursting is avoided by excreting the incoming water as fast as it enters. In the majority of cases the inflowing water is turned out in the form of a copious but very dilute urine, the kidney, antennal gland or other excretory organ taking on the function of water excretion in addition to its more primitive function of excreting waste products.

The excretion of a hypotonic urine, i.e. one which is hypotonic to the blood, means, of course, that animals possessing osmotic independence must possess excretory organs capable of separating the incoming water from the salts of the blood and actively excreting it; in other words, of doing osmotic work. In a number of cases, however, the urine is found to be isotonic with the blood. Here salts are not reabsorbed from the urine, but the loss through this channel is made good, partly by salts contained in the food and partly by the absorption of salts from the surrounding medium, usually by way of the gills. In either case, the maintenance of the osmotic gradient between the internal and external environments is only possible at the price of doing work, for living organisms, like non-living material, obey the laws of thermodynamics.

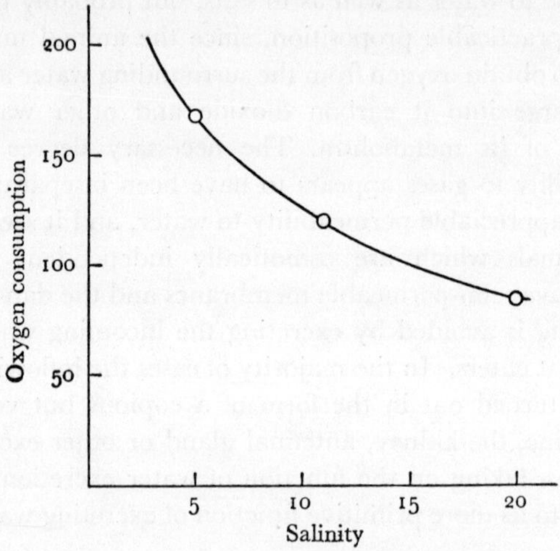

Fig. 2. Effect of external salinity on the oxygen consumption of *Carcinus*. Salinity in gm. NaCl per litre, oxygen consumption in arbitrary units.

If we measure the oxygen uptake of specimens of *Carcinus* at rest in sea water at various dilutions, we find that the rate of respiration increases with increasing dilution of the external medium, and this, of course, corresponds to increasing the osmotic gradient. The relation between oxygen uptake and external salinity is shown in Fig. 2 for *Carcinus*, and similar results have been obtained for other animals. It might be expected in view of these considerations that the respiratory rate of fresh-water animals in general would be greater than that of closely related marine species, since the latter have no need to withstand an osmotic gradient, and it has been shown that this is indeed the case for certain crustaceans (Amphipoda, Isopoda) (see table below). Other results of a similar nature could be quoted; they indicate one and all that the 'rateable value' of fresh water is much higher than that of the sea.

Species	cmm. O_2/gm./hr.
Marine:	
Gammarus locusta	207
Idotea neglecta	204
Fresh water:	
Gammarus pulex	404
Asellus aquaticus	645

In most animals we find that the influx of water is restricted to a fairly small part of the body surface, the rest of which is protected by the presence of a waterproof cuticle of some kind. In this way the rate of entry of water is much reduced and the excretory organ spared a great deal of osmotic work. This cuticle usually consists either of chitin, a complex polysaccharide found in the Crustacea, or else of keratin, a protein which is found in most vertebrates. The majority of fishes, for example, possess a waterproof coat made up of keratinous scales impregnated with a slimy mucin-like material. In one interesting case, that of the eel, the slime can be rubbed off merely by wiping the animal with a rough towel. If the eel is then returned to fresh water, water passes very much more rapidly into the body, the kidneys are unable to cope with the situation and the animal becomes water-logged and finally dies. This is an excellent demonstration of the osmotic as opposed to the purely defensive value of an impermeable cuticle.

The most commonly occurring arrangements for maintaining an osmotic gradient thus involve the following main factors:

(*a*) the possession of semi-permeable boundary membranes;

(*b*) the restriction of these to a relatively small area by the presence of an impermeable cuticle;

(*c*) the possession of an excretory organ capable of secreting a hypotonic urine, and therefore of doing osmotic work;

(*d*) utilisation of salts contained in the food; and

(*e*) the possession of special cells somewhere on the surface; these absorb salts from the environment and do the osmotic work which would otherwise fall upon the excretory organ.

This kind of mechanism is met with in a large number of fresh-water animals (cf. Fig. 3). Fresh-water Protozoa possess special contractile vacuoles which actively remove the water entering through the surface, while mosquito larvae living in fresh water receive water through the so-called 'anal gills' and excrete it again by way of the Malpighian tubules. In this and in some other cases it has been shown that active salt-absorption can take place also, even from fresh water. Such a process would clearly facilitate the maintenance of a high internal osmotic pressure. In the frog, however, the arrangement is somewhat different, for here the work of regulation appears to be carried out mainly by the skin, through which water enters, and not by the kidney, through which it leaves the body.

In the majority of fresh-water animals salts are actively reabsorbed from the urine, so that the latter is more or less hypotonic to the blood. The degree of hypotonicity is a rough measure of the rate at which water is forced into the body by the existing osmotic gradient. Now the excretion of a hypotonic urine requires appropriate specialisation of the excretory organ, just as, when (*e*) is involved, certain parts of the boundary membranes are

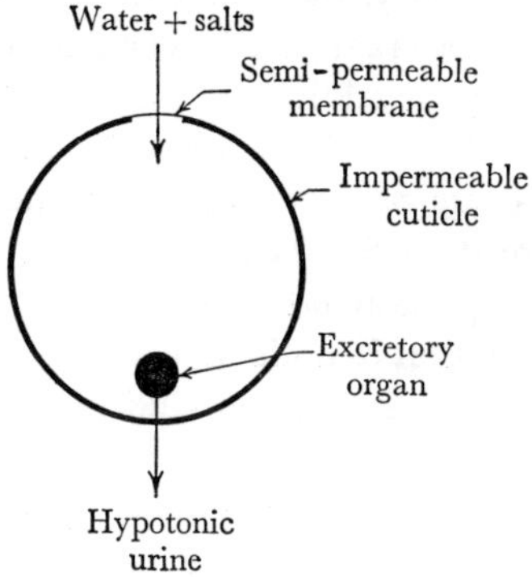

Fig. 3. Osmoregulatory mechanisms of a typical fresh-water organism.

specialised for salt absorption. Whereas the excretory organ of a marine animal, which is living in an approximately isotonic medium, need only filter off a protein-free blood filtrate, that of a fresh-water animal must in some way prevent loss of salts from the blood. The antennal glands of some Crustacea (Amphipoda) are of great interest from this point of view. In *Gammarus locusta*, a marine form, the tubule of this gland consists essentially of a filtration segment and a bladder segment, but in the closely related fresh-water form, *G. pulex*, the tubule is much longer. A new, salt-absorbing segment has been added. In *G. pulex* the salt-absorbing region reabsorbs salts from the blood filtrate produced in the filtration segment so that the urine finally contains less salts than the blood, to which it is therefore hypotonic. We can look upon the addition of this new salt-absorbing segment as an adaptation without which the maintenance of an osmotic gradient would in all probability have been impossible. Similarly, the tubule of the antennal gland of the fresh-water crayfish is longer than that of related marine forms by an additional segment. The kidney of the fresh-water fishes acts in the same kind of way. The glomerular filtrate passes down the tubule and salts are absorbed from it on the way.

One very important point remains to be mentioned. We have shown that osmotic independence requires that the animal shall possess some kind of control over the permeability of its surface membranes. This is particularly important in euryhaline animals, and the case of the euryhaline platyhelminth worm, *Gunda* (*Procerodes*), has been studied with extremely interesting results. *Gunda*, an estuarine animal, will live indefinitely in sea water and for considerable periods in stream water. It does not live indefinitely in the latter, however, for its surface is not ideally semi-permeable and salts are slowly lost from the body. Pantin found that *Gunda* survives better in some fresh waters than in others, poor survival being associated with a more rapid loss of salts from the body. The controlling factor was shown to be calcium. In hard waters the rate of loss of salts is much less than in soft waters, i.e. the permeability of the surface membranes is kept down by the presence of calcium. It is therefore believed that the calcium ion plays an important part in the regulation of membrane permeability, and in this connection one is reminded of the effect of calcium in preventing chilblains, which are themselves due to deranged permeability of the cell walls. From our present viewpoint the significance of the calcium effect is that a given animal might appear to be euryhaline or completely stenohaline according as calcium is or is not present in the surrounding water, a fact which has been confirmed in other animals besides *Gunda*. There is also a suggestion that in the migration of marine animals into fresh water, hard waters must probably have been more easily colonised than soft.

2. *Osmotic independence and evolution*

Although the conditions of life in sea water are such that marine animals have no need of osmotic independence, it is, as we have seen, an essential part of the stock-in-trade of estuarine and fresh-water animals. All fresh-water organisms are, in point of fact, osmotically independent. We may therefore reasonably infer that, if a given animal is osmotically independent, it or its ancestors must at some time or other have lived under estuarine conditions.

Now if we compare the values of Δ_i and Δ_e for a number of different animals, we discover the striking fact that, whereas some invertebrates are osmotically independent and others not, *all the vertebrates are capable of maintaining an osmotic gradient.* This is shown diagrammatically in Fig. 4. Even in the special and very peculiar case of the elasmobranchs (cartilaginous fishes) the salt content of the blood is about the same as that of the teleosts (bony fishes) and the rest of the vertebrates; the high total osmotic pressure of elasmobranch blood is largely due to the fact that it contains a large amount of urea. In fresh-water forms there is about 0·6 and in marine species 2–2·5 per cent of urea, the significance of which we shall discuss presently.

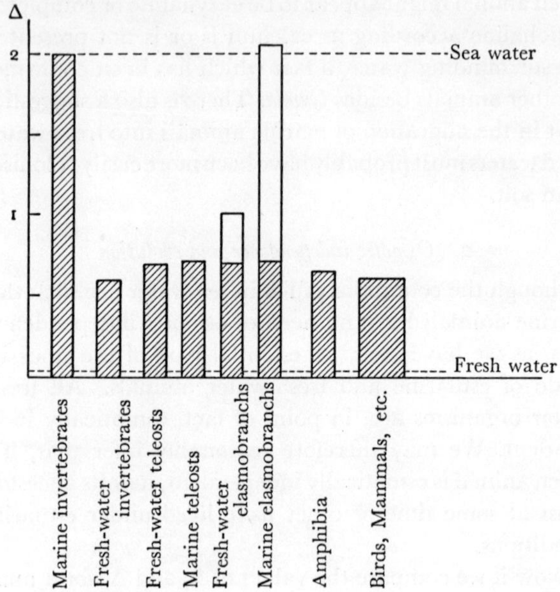

Fig. 4. Osmotic pressures of bloods of various animals compared with those of fresh and sea waters.

On account of their many morphological resemblances it is universally held that the vertebrates of today are all descended from a common ancestral stock, and from the fact that they are all osmotically independent we may conclude that this ancestral stock must at one time have lived in estuaries. It was formerly held that the vertebrates originated in the sea, but this physicochemical evidence of an estuarine or fresh-water ancestry has found other support in geological evidence, and a closer study of the processes of osmoregulation in the fishes serves to confirm this point of view.

1. *Fresh-water teleosts.* Here Δ_i is greater than Δ_e, and water therefore passes into the fish across the semi-permeable membranes of the gills and the mucous membranes of the mouth. The incoming water dilutes the blood but is filtered off by the glomeruli of the kidney in the manner we have described, the salts being reabsorbed from the glomerular filtrate as it passes down the kidney tubules, so that a copious stream of very dilute urine is produced. The maintenance of the high Δ_i is aided by active salt-absorption by special cells in the gills. This mechanism is illustrated in Fig. 5*a*; it is essentially the same as that shown in Fig. 3 and is common to many fresh-water organisms. Any loss of salts that may take place can be made good by the use of food salts or by the absorption of salts from the surrounding water.

2. *Marine teleosts.* Here Δ_i is less than Δ_e and water passes out of the body, so that whereas the danger facing the fresh-water teleost is that of death by flooding, the marine teleost must constantly fight against the danger of death from dehydration. It must contrive in some way to obtain water from the surrounding sea, and this involves separating it from the salts which are also present. Three mechanisms seem possible at first sight. The fish might absorb water (but not salts) in some way, perhaps through the gill membranes, but apparently it has not evolved the power to do this. It might swallow sea water and absorb it from the gut, but in this case it would be

difficult to avoid absorbing the unwanted salts as well, since the gut membranes must, after all, be permeable to the molecules of sugars, amino-acids and the like produced by digestion of the food. There remains the possibility of absorbing both water and salts and excreting the salts again in the form of a hypertonic urine or otherwise.

Now it is known that the marine teleost obtains water by swallowing, for if swallowing is prevented by introducing a rubber balloon into the oesophagus and inflating it, the fish is unable to keep up its regulation and soon dies of dehydration. It can be shown that salts and water are both absorbed from the gut of the normal fish, yet the urine, far from being hypertonic to the blood, is actually slightly hypotonic. It seems that the fish kidney has not evolved the power of secreting a hypertonic urine, and we shall see why in a later section. The unwanted salts are excreted, not by the kidney, but by special cells, the so-called 'chloride secretory cells', situated in the gills. These are reminiscent of the salt-absorbing cells found in certain fresh-water animals (cf. p. 32), for here, though they work to remove salts *from* the animal, there is again a transport of ions from a medium poor in salts to one which is relatively richer.

The marine teleost, then, obtains water by swallowing sea water and absorbing it, salts and all, from the gut. The unwanted salts are removed by means of special secretory cells in the gill membranes, and the water which remains makes good the loss due to osmosis. This mechanism is shown in Fig. 5*b*, but it should be mentioned that in migratory fishes such as the eel and the salmon, both types of regulatory mechanism are present and brought into action according to circumstances.

Water is thus a valuable commodity for the marine teleost, and we accordingly find that not much is wasted in the formation of urine. The urine is as concentrated as it can be made, that is to say, nearly isotonic with the blood, and the difference in the nature of the urinary secretion is reflected in the structure of the kidney. It is believed that the glomerulus was developed in response

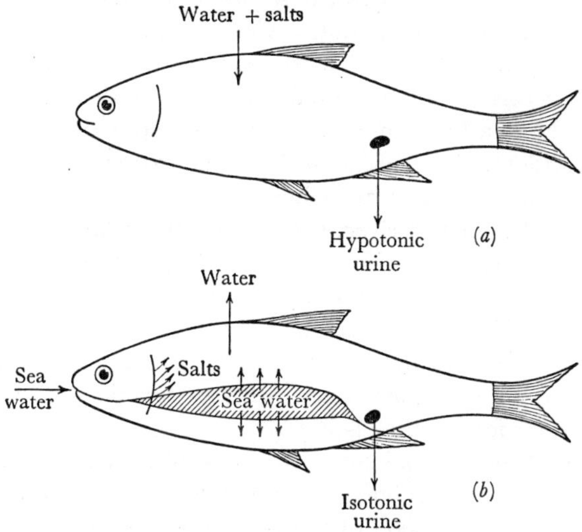

Fig. 5. Osmoregulation in teleostean fishes. *a*, fresh-water species. *b*, marine species.

to the necessity of filtering off large amounts of water, as in the fresh-water teleosts, the kidneys of which are abundantly supplied with glomeruli. In the marine teleosts, however, the possession of glomeruli would be more of an embarrassment than an advantage on account of the necessity of conserving water, and few glomeruli are present, the whole kidney giving the histological appearance of having at one time possessed glomeruli which later degenerated. Indeed the embryonic kidneys of some marine teleosts possess numerous glomeruli but these degenerate in the course of development, suggesting that these species had a fresh-water origin.

The *migratory fishes* present an interesting problem. As has been pointed out, they possess mechanisms which allow them to maintain a virtually constant internal salinity in fresh, estuarine and sea water alike. Glomeruli are necessary while the animals are in fresh water but would be a hindrance rather than a help under marine conditions. It seems unlikely in the extreme that new glomeruli can develop when an eel or a salmon, for example, enters a river, and degenerate again when the animal goes to sea. How the change-over is accomplished we do not know; possibly the blood supply to the glomeruli can be cut off, perhaps by constriction of the renal artery and/or vein when the animals are in sea water, so that the kidney becomes functionally aglomerular for the time being.

3. *Marine elasmobranchs*. The blood of a marine elasmobranch contains about the same amount of salts as that present in the blood of the marine teleosts (see Fig. 4), but the presence of 2 per cent or more of urea brings up the total osmotic pressure to a level slightly higher than that of sea water. The resulting small osmotic gradient tends to drive water into the fish, which therefore resembles a fresh-water rather than a marine teleost. By conserving urea instead of excreting it, the marine elasmobranch evades the necessity of developing a taste for sea water and also that of developing special secretory

cells with which to eliminate unwanted salts. The escape of urea from the fish is prevented by having gills which are impermeable to urea, while the control of its loss in the urine is accomplished by a special segment of the renal tubule which, incidentally, is glomerular. This urea-absorbing segment is peculiar to the elasmobranch fishes, and its function in reabsorbing urea from the glomerular filtrate is analogous to that of the salt-absorbing segment of the fresh-water teleosts.

Since the osmotic gradient driving water into the fish is quite small, water enters fairly slowly and the urine, though hypotonic, is neither so weak nor so abundant as that of fresh-water teleosts.

4. *Fresh-water elasmobranchs.* Although the majority of modern elasmobranch fishes are marine, a number of fresh-water species are known, and of these the sawfish, *Pristis*, is a well-known example. The salt content of the bloods of the fresh-water forms is greater than that of fresh water itself, so that even if no urea were present at all there would still be a considerable osmotic gradient to drive water into the body. Urea is nevertheless present, and since it only serves to increase the rate at which water comes in, and therefore the amount of work imposed upon the kidney, it is believed that the fresh-water elasmobranchs are forms which at one time lived in the sea and later entered the rivers, carrying with them a legacy from the past in the form of a well-marked uraemia. Had they retained the 2 per cent or more of urea which is characteristic of the marine elasmobranchs, water would enter these fresh-water forms very rapidly indeed, and, apparently as an adaptation to this, we find that their urea content has been reduced to about

0·6 per cent. The osmotic pressure of the blood is therefore well above that of the fresh-water teleosts, so that the rate of entry of water is correspondingly greater and the kidney must work harder and produce a still more dilute and even more copious urine than that of the fresh-water teleosts. It would clearly be to the animal's advantage to get rid of urea altogether and thus economise on osmotic work, but apparently during their long stay in the sea the tissues became so accustomed to the severe uraemia which was so advantageous then, that the heart of the fresh-water elasmobranch of to-day is unable to beat in the absence of urea.

Summing up the case of the fishes, it seems that we can say that they all originated in fresh water. The presence of a water-excreting device in the form of the glomerulus is evidence of an estuarine or fresh-water ancestry. The glomerulus degenerated in those teleosts which went to sea, in response to the necessity for conserving water as much as possible, but it persisted in the elasmobranchs, for these forms found in a physiological uraemia an alternative method of protecting themselves against the loss of water. Some of these forms have returned to fresh water, and though they still show the phenomenon of uraemia for other reasons, the degree of uraemia has been greatly reduced in response to the change in the osmotic conditions to which they are exposed.

CHAPTER III

1. *The colonisation of dry land*

IN spite of the instability of its climatic conditions the dry land offered many advantages to such animals as could adjust themselves to a terrestrial environment. The possibility of rapid movement, a copious supply of oxygen, an abundance of vegetation for food and shelter, and protection from aquatic enemies were a few of the advantages to be gained. The earliest settlers must presumably have been herbivorous for the most part, but with an almost unlimited supply of food at their disposal the herbivores must soon have multiplied sufficiently to make terrestrial life profitable for carnivorous forms as well. But there were many difficulties, some of which have had important biochemical consequences.

Perhaps the greatest biochemical factor was the fact that, on taking to dry land, the animals became cut off from the abundant supply of water which constituted their former aquatic environment. Henceforth they were faced with the ever-present danger of desiccation, and the conservation of that all-important biochemical reagent, water, became a matter of the most urgent necessity. It was necessary also that the respiratory organs should be so modified as to allow of air-breathing, and the fact that the respiratory epithelia were no longer surrounded by water made it imperative that the excretion of certain substances such as ammonia, formerly disposed of across the gill membranes and other parts of the body surface, should be carried out in future by suitably specialised excretory organs. But ammonia is a highly toxic substance and its rapid and efficient elimination by the excretory apparatus would call for a copious supply of water. Hence, as we shall see, adaptation to terrestrial life has been attended by important changes in the mechanisms of nitrogenous metabolism.

The general speeding up of life which became possible in a medium containing abundant oxygen and of negligible viscosity called for increased specialisation of the central nervous system. In the insects, the size of which is limited by other considerations, the evolution of a large and complex brain was precluded for reasons of space, and here specialisation proceeded in the direction of complex instinctive behaviour, but in the vertebrates there arose the specialised brain and the complex system of hormonic control which have reached their highest development in the mammals. Step by step more complex reflexes were established, more subtle adjustments were called for, memory, thought and speech were evolved and finally, as the crowning achievement so far, there appeared the ability to record thought and experience.

Before going on to consider the mechanisms which have been evolved for dealing with some of the difficulties inherent in terrestrial life, it will be interesting to review very briefly the paths which have been followed by animals in migrating from the sea to the land. The journey might be accomplished either directly by migration up the littoral zone or indirectly by way of fresh water. The littoral zone is liable to severe mechanical disturbances due to wave action as well as to the usual climatic conditions. There are rhythmic changes of temperature, wetness and salinity, but this zone is nevertheless densely populated wherever a foothold can be found or burrowing is possible. The animals spend a part of their time in water and part in air, so that there is every reason for them to become adapted for aerial respiration. But on the whole it seems likely that the route through fresh water is the one that has been most frequently followed. In shallow pools, marshes, swamps and the like the oxygen content of the water may, and frequently does, fall to a very low level, and there is therefore a tendency for animals inhabiting such regions to become adapted for aerial respiration while their mode of life is still aquatic. In certain tropical swamps the oxygen content of the water is always extremely low and usually negligible, and many of the fishes living there are adapted for air breathing. Carter has summed up the respective merits of fresh and salt water as starting points for landward migration in the following words:

In the fresh waters the various changes necessary for the migration towards terrestrial conditions have been induced in the fauna one after another. Lack of oxygen produced aerial respiration while the animal was still aquatic; this made aestivation, in response to occasional droughts, more easy; and aestivation, since it produced adaptations against desiccation, etc., led to life in amphibious conditions, and so to truly terrestrial conditions....On the shore the sea water is well oxygenated and there is no impulse for an animal to evolve aerial respiration until it has become active in the air. There, the many changes which go to make up the difference between aquatic and terrestrial conditions occur together. There are no intermediate environments to introduce the animal to the changes in the environmental conditions serially.

Even so, many terrestrial snails and the land crabs appear to have migrated up the marine littoral instead of by the more usual estuarine → freshwater → land route.

In addition to adaptations which concern the adult animal, however, there are numerous conditions which affect its development. The terrestrial embryo must be provided with water as well as with the food and salts required by fresh-water embryos. This has given rise to two most important adaptations. In some cases, e.g. the birds and most of the reptiles, the egg is provided with water when laid and protected against its loss by being deposited in a damp microclimate of some kind, or by being surrounded with a more or less impermeable shell or membrane of some kind, and eggs of this latter kind are said to be 'cleidoic' (from κλειδόω, I enclose). An alternative method was that adopted by the mammals, which became viviparous. In this case the developing embryo could be carried about inside its mother and provided with water at the mother's expense. Viviparity is probably better than cleidoicity since instead of having to make a nest and incubate the eggs, the mammalian mother could carry her eggs about with her so that her range of movement in search of food was accordingly less restricted. Cleidoicity and viviparity alike demanded intrauterine fertilisation, which is therefore to be regarded as yet another adaptation to life under terrestrial conditions.

It is interesting to notice in passing that eggs can only be laid away from the water provided that the young do not hatch in a larval form. Notable exceptions to this general rule are, of course, common enough among insects, but broadly speaking larval stages must be compressed into the pre-hatching period if this has not already been done. Thus the common frog, *Rana*, lays aquatic, non-cleidoic eggs, the young hatching as tadpoles which metamorphose later on, but there are some frogs, e.g. *Hylodes*, a West Indian tree frog, which lay terrestrial eggs, and here the tadpole stage is passed through inside the egg, the animal hatching as a perfect but miniature frog.

The variability of the temperature of terrestrial environments has played a part of considerable importance in the evolution of reproductive processes. At certain periods, notably at the time of gastrulation, the developing embryo is particularly susceptible to thermal and other changes. Perhaps the early (Mesozoic) reptiles became extinct because their eggs were set to develop at the relatively high temperatures of Mesozoic times and proved unable to withstand the cooler conditions which some authors believe to have accompanied the passing of the Mesozoic and the advent of the Cenozoic. It has been suggested, interestingly if rather speculatively, that another reason for the extinction of the reptiles was that the ferns which had previously been the dominant form of plant life were being replaced by flowering plants, thanks largely to the fertilising activities of the insects which began to be abundant about the same time. This then meant a change in reptilian diet which, when one remembers the purgative action of fern oils, may be thought to have caused many of the animals to die of constipation!

Perhaps the only safe way of steering the embryo through the difficult period of gastrulation must have been to see that it developed at a constant temperature. Some modern snakes protect their eggs by wrapping their coils round them, while others lay the eggs in dung heaps or piles of decaying rubbish where they will be kept gently warmed. But the most satisfactory way of dealing with the problem seems to have been that of becoming homoiothermic, the eggs being incubated either inside or outside a body of which the temperature was very accurately regulated. This, perhaps, is how the birds and the mammals secured and maintained their supremacy over the reptiles; with Graham Lusk, we 'may recall seeing a cat basking in sunshine in zero weather, and contrast its life with that of the alligator, which would freeze under like conditions'. The reptiles, which are not homoiothermic, are necessarily restricted for the most part to climates hot enough to allow their eggs to develop properly, but birds and mammals are to be found the world over.

The evolution of homoiothermism has been achieved through the development of a complex series of mechanisms for controlling the metabolic rate, sweating, shivering and so on. Homoiothermism involves a loss of efficiency in the thermodynamic sense because so much energy is dissipated in the form of heat but, by adding yet another to the number of constant properties of the internal environment, it has made possible still further complexity and flexibility of central nervous control.

2. *The conservation of water*

We have already seen that one of the conditions of life in fresh water is that most of the surface of the body shall be covered by an almost impermeable cuticle. In this respect the fresh-water animal would have the advantage over the marine form on migrating to land. All that would be necessary would be final and complete waterproofing of the already relatively impermeable cuticle. The chitinous covering of the Arthropoda has been successfully used as a water-retaining coat by numerous terrestrial forms such as the land crabs and the insects. Waterproofing in the insects has been achieved by impregnation of the cuticle with the curious 'insect waxes', together with a tanning process. There is an interesting series of isopod crustaceans in which the waterproofing of the cuticle has been accomplished with progressively greater success. *Asellus* is an aquatic form which never leaves the water, while *Ligia*, though amphibious, cannot go far from water as it is not well protected against its loss. *Oniscus*, a terrestrial form, does not require periodic immersion as does *Ligia*, but nevertheless can only live in rather damp surroundings, while *Armadillidium*, the common wood louse, is relatively resistant and can live for some time in dry air.

In all these forms the cuticle is chitinous in nature, but there is little doubt that the keratinous coat of the vertebrates far surpasses the crustacean chitin in general utility. It is not only waterproof, but very flexible and tough, and from it are derived the scales of fishes, the skin of frogs, snakes and lizards, the armour-plating of tortoises and crocodiles, as well as the hair and feathers of the higher vertebrates. According to James Gray, 'the origin of truly terrestrial vertebrates was associated with the development of a skin which is relatively impermeable to water. The skins of fishes and most amphibia are freely permeable to water, and water is readily lost when the animals are exposed to ordinary atmospheric conditions. The skin of the lizard, on the other hand, is relatively impermeable to water and the animal is therefore able to withstand exposure to air without serious consequences.' Thus, to take Gray's own rather vivid example, a skinned lizard loses water to the air at about the same rate as a normal newt, and the rates of loss are about the same whether the animals are alive or dead.

There are, however, reasons for believing that the permeability of fish skins to water is relatively small (see pp. 37–42), at any rate so long as the animal is immersed in water. We may suspect that the apparent 'free permeability', to which Gray alludes, when the fish is exposed to air, is due to the drying of the mucous material with which the keratinous skin is impregnated. Perhaps the essential difference between the skin of the lizard and that of the fish consists in the replacement of a mucous by a cement-like material which remains waterproof even when exposed to air and is itself exposed to desiccation; something, in other words, analogous to the insect waxes. But the cuticle can only afford passive protection against the loss of water, and active measures had also to be taken to acquire water and to prevent its loss in the excreta. 'Reptiles', wrote Gray, 'are the earliest type of vertebrate which drinks water through the mouth and absorbs it by the alimentary canal. Typical amphibia, on the other hand, do not drink: they imbibe water over the whole surface of their bodies.' But there are differences even among amphibians, for toad skins are usually less permeable to water than those of frogs.

As far as the loss of water in the excreta is concerned, it will be remembered that the glomerulus of the kidney of the marine teleosts has degenerated in response to the necessity of conserving water, and the same degenerative process has taken place in the kidneys of certain terrestrial animals, notably the dry-living reptiles (snakes, lizards), whereas in the amphibious reptiles (turtles) the glomerulus persists. Similarly, while the common frog possesses a glomerular kidney, that of a desert frog, *Chiroleptes*, is believed to be aglomerular. The birds and mammals, however, have protected themselves to some extent by reducing the size of the glomerulus, but in the main by adding a water-absorbing segment, the loop of Henle, to the kidney tubule. The loop of Henle reabsorbs water from the glomerular filtrate and makes possible the formation of a very concentrated urine, a feat which cannot be accomplished by kidneys in which no such loop is present.

In the birds the action of the loop of Henle is reinforced by yet another device, one which is found in some other terrestrial animals as well. The rectum and the ureter or urethra are fused together to form a cloaca. The kidney must always allow enough water to pass to prevent its ducts and tubules being blocked by solid masses of excretory substances, but by allowing the still liquid urine to pass into the rectum instead of being discharged to the exterior, advantage is taken of the power of the rectum to reabsorb water, and a semi-solid urine like that of the birds is formed. Closely comparable arrangements are found in all classes of reptiles. Essentially similar to this is the arrangement used by many insects. Here a liquid urine is secreted by the cells of the Malpighian tubules, some of the water being absorbed as it passes down into the rectum. The cells of the so-called 'rectal glands', as Wigglesworth has shown, then complete the reabsorption of excess water and a semi-solid or pellet-like mass is excreted. Even where no recognisable rectal glands are present the hind-gut is commonly lined with cells having the same histological characteristics and which may reasonably be suspected of discharging the same function.

The mechanisms for protection against loss of water are thus of two kinds; passive protection is provided by a waterproof cuticle and active protection by the retention or reabsorption of water which would otherwise be lost in the excreta.

External sources of water are not the only ones we have to consider, however. Metabolic water is continually produced within the animal as a result of the oxidation of food substances. The following figures show the average amounts of water formed during the complete oxidation of 100 gm. of protein, carbohydrate and fat respectively, and they show that the oxidation of fat provides a particularly rich source of metabolic water:

	Water (gm.)
Protein	41·3
Carbohydrate	55·5
Fat	107·1

It is therefore extremely interesting to find that there is often a strong emphasis on the oxidation of fat in animals living under conditions of acute water shortage. Of the stored foodstuffs metabolised by the starving mealworm, for instance, the proportions of fat: carbohydrate: protein are approximately 8:2·5:1, while in the hibernating marmot and hedgehog the respiratory quotient has a value of 0·7, indicating that fat and practically nothing else is being burnt. The ability of the camel to travel for days without a drink is due, not to storage of water in the hump, but to a heavy emphasis on fat metabolism. Similarly, fat accounts for over 90 per cent of the total foodstuffs metabolised during the development of the chick embryo in its 'closed box' egg. In each case we find that fat metabolism is strongly stressed, with consequent production of large amounts of metabolic water, as an adaptation to life with a limited water supply. This is all the more interesting because, in spite of the intense metabolism of fatty acids, no ketosis is demonstrable.

As will be noticed from the figures just quoted, there is at the same time a marked tendency to spare the metabolism of protein, and this must probably be associated with the difficulty of disposing of nitrogenous waste products when the water supply is restricted. We have already alluded to this aspect of terrestrial life and must now discuss it at greater length.

CHAPTER IV

1. *The excretion of nitrogen*

The chief source of nitrogen is the α-amino-N of the amino-acids from which the proteins of the food are built up. These amino-acids are liberated by the action of proteolytic digestive enzymes and subsequently, for the most part, deaminated with the production of ammonia. About 90 per cent of the total nitrogen excreted by animals originates in this way.

The following representative figures show the percentages of the total nitrogen excreted in various forms by invertebrates; they were obtained by averaging the results secured by analysing the excreta of seven different aquatic species:

	% of total N
Ammonia	52·2
Urea	6·1
Uric acid	1·3
Amino-acids	14·0
Undetermined	26·4

These values, which may be taken as fairly typical for aquatic invertebrates, show that while a considerable proportion of the total nitrogen consists of unchanged amino-acids, the principal end-product is ammonia. The 'undetermined' N includes small amounts of various compounds such as purines and betaines, with which we shall deal presently. The excretion of amino-acids in amounts which, in individual cases, may be as high as 30 per cent of the total N suggests that the metabolic machinery of the invertebrates may be deficient in some way, and incapable of dealing successfully with all the amino-acids derived from the food. On the other hand, it is quite possible that they may, partly at any rate, simply have leaked out across the body surface, which we know to be permeable enough in the case of the marine forms to allow salts to pass through.

The following series of data was obtained by analysis of the urine of a marine teleost, the angler fish, *Lophius*, and shows some interesting features:

	% of total N
Ammonia	56·0
Urea	5·7
Uric acid	0·2
Amino-acids	0·0
Creatine	6·5
Trimethylamine oxide	28·2
Undetermined	3·4

Here we can account for practically all of the nitrogen, and again we find that the principal end-product is ammonia. Small amounts of creatine are present, and creatine or creatinine or both are always to be found in the excreta of the vertebrates. Amino-acids are absent, indicating, perhaps, a metabolic superiority over the invertebrates, but what is particularly striking is the presence of nearly 30 per cent of the total N in the form of trimethylamine oxide. We shall return presently to this point.

Leaving aside the elasmobranch fishes, which, as we know, had ulterior motives in converting their waste nitrogen into urea, we can say that *it is characteristic of aquatic organisms that their major nitrogenous end-product is ammonia.* This compound requires a plentiful supply of water for its efficient excretion; it is extremely toxic and must be removed rapidly and efficiently from the body. Ammonia is such a familiar household article that its extreme toxicity is not, perhaps, as widely appreciated as it ought to be. Sumner showed that if pure, crystalline urease is injected into rabbits, ammonia is formed by hydrolysis of the small amounts of urea present in the blood. Death ensues when the concentration of ammonia reaches about 1 part in 20,000. This is too small to produce any significant change in the pH of the blood, so that death cannot be attributed to alkalaemia. Nor is it attributable to any specific toxicity of the enzyme-protein, for similar experiments have been carried out on hens, the blood of which contains no urea, without untoward results. There can be little if any doubt that Sumner's rabbits died from ammonia poisoning. Toxic though it is, however, ammonia can be easily disposed of by aquatic organisms, since it is a readily diffusible substance and can escape rapidly into the surrounding water.

But when we come to the Amphibia, which spend part of their time on land and part in the water, we find that the principal end-product is no longer ammonia but urea. The common frog hatches from the egg as an aquatic, ammonia-excreting tadpole, but at about the time that it develops the ability to go on to the land, ammonia excretion gradually ceases and is superseded by that of urea. This is evidently an adaptation without which it would be impossible for the animals to spend even a part of their time on the land on account of the difficulty of getting rid of ammonia. Urea, like the ammonia which it replaces, is a very soluble and very diffusible substance, but differs from it in that it is non-toxic even in relatively high concentrations. It could therefore be formed and remain in the body for some time before being excreted, a fact of which these amphibious animals have taken full advantage.

There are, however, some amphibia which, like *Xenopus*, have made a secondary return to water, and are now essentially aquatic. *Xenopus* and a number of similar species still excrete urea, as evidence of their amphibian ancestry, but the main nitrogenous end-product has reverted to ammonia. However, aquatic though they are, these animals can survive removal from water into damp air for some days but then produce little ammonia but much urea, and can store the latter in their blood and tissues until they are returned to water.

We must thus regard urea production as an adaptation to conditions of water shortage. But urea is produced as the main nitrogenous end-product of the elasmobranch fishes, although the latter are certainly not amphibious in any sense of the word. However, this apparent anomaly can be explained fairly readily if we compare the elasmobranchs with the teleostean fishes. It has been pointed out that the teleosts excrete a good deal of waste nitrogen in the form of trimethylamine oxide, and the significant point for the present argument is that this compound is only produced by those teleosts which are marine. Now it will be remembered that whereas an abundant supply of water is continually being forced into the body of the fresh-water teleost, the marine forms must constantly fight against the danger of desiccation—water is hard to come by and the supply of it is severely limited. May we not therefore suspect that the production of trimethylamine oxide by the marine teleosts is in reality an adaptation to this shortage of water, in that it replaces the toxic substance ammonia by a soluble, non-toxic and practically neutral end-product? The absence of trimethylamine oxide from fresh-water forms would thus be intelligible, for there is in their case no need to 'mask' ammonia since an almost unlimited water supply is available for its excretion. Following up this line of argument, it may be suggested that the production of urea by the marine elasmobranchs originated as an adaptation to a similar limitation of the water supply. The fact that their gills became impermeable to this new end-product meant that by its accumulation in the blood the osmotic gradient could be diminished and the battle for water made less severe, so that the adaptation of the tissues to an increasingly severe uraemia eventually relieved these animals of all anxiety regarding their water supply.

It has been suggested that the trimethylamine oxide of the marine teleosts may perhaps be of some osmo-regulatory significance, but the fact that it appears to diffuse readily across the gills argues against this, since the osmotic pressure of a dissolved substance cannot be exerted at a membrane through which it can pass freely. In any case the amounts of trimethylamine oxide found in the blood and tissues as compared with the urine are relatively small, so that if this compound has any osmo-regulatory value at all it must be very slight at most. In any case it is arguable that trimethylamine oxide is not produced by the fishes themselves but arises from the food. Many marine organisms that serve as fish food certainly do contain this curious substance.

Passing on now to consider truly terrestrial groups,

such as the snakes, lizards, birds and mammals, we might expect to find urea still in favour as an end-product of nitrogenous metabolism. But here we find a parting of the ways, the mammals continuing to excrete urea for the most part, while the birds and reptiles have abandoned it in favour of uric acid. The metabolism is said to be *ureotelic* in cases where urea is the main end-product, and *uricotelic* where the principal product is uric acid.

This difference among the higher vertebrates has been shown by Needham to be correlated with the mode of reproduction. Ureotelic metabolism is associated with viviparity, uricotelic metabolism with development within a cleidoic egg. The following reasons may be advanced to account for these facts. Cleidoic eggs develop under a very special set of conditions, the most notable of which is strict limitation of the water supply. The developing embryo, like any other living tissue, metabolises fat, carbohydrate and protein in the course of its development, producing carbon dioxide, water and ammonia. The first of these can readily be disposed of, while the second is valuable to any terrestrial organism. The third, ammonia, is toxic to the organism and cannot therefore be allowed to remain as such in the tissues. If it were turned into urea its toxicity would be overcome, but by the end of development the amount of urea produced would probably be sufficient to cause considerable uraemia, and thus damage the cells by disturbing their osmotic relationships. Apparently the elasmobranchs (and at least one species of frog according to a recent report) are the only animals which can withstand more than the mildest uraemia. It was necessary therefore that the embryo should convert its waste ammonia not into urea but into some substance which was at the same time non-toxic and sufficiently insoluble to be precipitated and so exert no harmful osmotic effects. Such a substance was found in uric acid and the production of urea has been abandoned by the birds, snakes and lizards in favour of that of uric acid. Attached to the belly of the developing chick embryo is a little membranous bag, the allantois, which serves as a dump heap for the uric acid produced during embryonic life, and at the end of development its slimy contents include numerous nodules of solid uric acid.

The production of uric acid was not necessary in the case of the mammals. Being viviparous, they are in connection with the maternal blood stream throughout embryonic life, and the water supply is therefore much less restricted. The formation of urea suffices to detoxicate ammonia produced during the metabolism of proteins, and urea can readily diffuse away across the placenta to be excreted by the maternal kidneys.

Similar considerations may apply to some extent to amphibious animals. Most Amphibia lay their eggs in water, so that there is no obvious reason why ammonia should not be formed as it is in other non-cleidoic eggs, and, in point of fact, the ureotelic metabolism of the frog is only fully established at metamorphosis. Tortoises and turtles frequently lay their eggs in mud or wet sand. Under these circumstances diffusion of ammonia from the eggs might well be rather a slow process and the replacement of ammonia by urea or uric acid would probably prove advantageous. It has been reported, however, that the egg shell of some members of this group is even less permeable to water than that of a typical bird. Moreover, some species lay their eggs in dry rather than damp situations, in which case extreme impermeability would be probably advantageous to the embryo. Unfortunately the excreta of only a few of these reptiles have been analysed, and while wholly aquatic species are ammoniotelic, amphibious forms are ureotelic. Of the dry-living species, some are ureotelic but others excrete urea and uric acid together. Finally in at least one desert-living species uric acid accounts for practically the whole of the non-protein nitrogen excreted. It is evident, therefore, that in the tortoises and turtles we have an intermediate group, lying between the ureotelic Amphibia and the uricotelic snakes and lizards: perhaps the uricotelism of the more progressive of these reptiles is still being evolved in the slow but steady manner usually associated with tortoises.

The type of nitrogenous excretion developed in response to the conditions of embryonic life persists when the animal becomes adult, and we can therefore say that, *among truly terrestrial animals, ureotelic metabolism is correlated with viviparity, uricotelic metabolism with cleidoicity of the egg.*

The elasmobranch fishes had a special difficulty to face in that the young ones must be provided with a supply of urea with which to keep up the osmotic pressure of the blood. In some forms this is done by laying urea-proof eggs. Ammonia produced during the period of development is turned into urea, and this, being unable to escape, accumulates and is available when the young fish hatches out. Other elasmobranchs achieve the same result by viviparity, the necessary supply of urea being obtained in this case from the maternal organism.

Up to the present we have only dealt with the nitrogenous metabolism of animals belonging to the Vertebrata, and the main points of importance are summarised in Table 5. Although we cannot stop here to go into details, it must be mentioned that the correlation between uricotelic metabolism and a cleidoic egg among terrestrial animals is not confined to the vertebrates. The other two groups which have been predominantly successful on dry land are the gastropods and the insects, the latter being almost exclusively terrestrial. In neither case are the eggs very well protected against the loss of water by evaporation, but usually they are laid in damp situations, at the roots of grasses or in other moist microclimates of various kinds. In both cases we find that the eggs approximate to the cleidoic type, and that the main nitrogenous waste product is again uric acid. Ureotelism seems seldom, if ever, to have been exploited by the invertebrates. Unfortunately research here is so backward that it would be out of place to discuss it at any length in a book of this kind. It might be mentioned, however, that just as the elasmobranchs found a good use for their waste nitrogen, so some of the insects have contrived to turn waste nitrogen to a useful purpose, as was first shown by Gowland Hopkins. The white wing pigment, leucopterine, of certain butterflies, and the bright yellow pigment, xanthopterine, of the wasp, are two examples of a group of pigements known as pterines which are closely similar in chemical structure to the uric acid group (purines).

Table 5

	Habitat	Water supply	G	H	Principal N end-product	Egg habit
Pisces:						
Elasmobranchii	SW	Good	+	−[1]	Urea	Urea-proof or viviparous
	FW	Good	+	−[1]	Urea	
Teleostei	FW	Good	+	−	Ammonia	Non-cleidoic
	SW	Poor	−	−	Ammonia[2]	Non-cleidoic
Amphibia	A	Poor	+	−	Urea	Non-cleidoic
	FW	Good	+	−	Ammonia[3]	
Reptilia:						
Turtles	A	Poor	+	−	Urea ± uric acid	Cleidoic?
Snakes, lizards	T	Bad	−	−	Uric acid	Cleidoic
Aves	T	Bad	+	+	Uric acid	Cleidoic
Mammalia	T	Bad	+	+	Urea	Viviparous

FW = fresh water.
SW = sea water.
A = amphibious.
T = terrestrial.

G = glomerulus.
H = loop of Henle.

[1] Urea-absorbing segment.
[2] Plus some trimethylamine oxide.
[3] Plus some urea.

2. *Recapitulation*

Before leaving the question of eggs and evolution, some mention must be made of the much-debated theory of recapitulation, according to which an embryo in the course of its development passes through stages at which

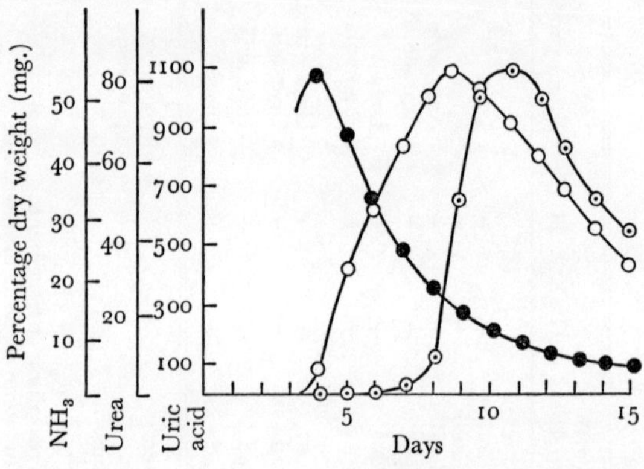

Fig. 6. Nitrogen excretion of chick embryo (after Needham). ● Ammonia; ○ Urea; ⊙ Uric acid.

it resembles to a greater or lesser extent the embryos of the animals from which it has evolved. In other words, the developing embryo is a kind of museum of evolution. The mammals, for instance, had fish-like ancestors, and early in its development the mammalian embryo actually possesses gill slits like those of a fish, while just before birth the body of the human embryo is covered by a coat of fine hair, very reminiscent of the 'monkey coat' of its less remote ancestors. Even after birth a human child shows signs of its simian affinities, for weak though it is in almost every respect, it is capable of supporting the entire weight of its body by the grip of one hand. In the same sort of way the early chick embryo has a simple tubular heart like that of a fish, and many other instances could be cited.

It would be disappointing if we could not find some chemical relics in this evolutionary museum, and one example of chemical recapitulation has already been cited in connection with development of ureotelism in the frog (p. 56). It has been shown by Needham that the chick embryo also appears to recapitulate in a chemical sense. At first it behaves like an aquatic organism and excretes ammonia, but after a few days is found to be excreting urea like an amphibian, and finally the ureotelic is abandoned in favour of the uricotelic type of excretion. Needham's curves, which are reproduced in Fig. 6, show this very beautifully and are worth close attention: attention must be given to the different scales used in the graph. But in this case, unhappily, it now appears that there are three distinct and separate activities with no continuity comparable with the smooth and continuous turnover from ammonia to urea taking place, for example, in a developing tadpole.

The theory of recapitulation, then, gets a little support from the chemical evidence, but not enough to transform it from a slightly disreputable hypothesis into a soundly established theory.

3. *The metabolism of nitrogen*

As has been mentioned, the principal source of nitrogen is the protein of the food. The nitrogen is liberated in the form of ammonia from the amino-acids set free by proteolytic enzymes, and thus far the metabolism of nitrogen follows the same course in all animals. A certain amount of the amino-acids goes to the repair of tissues and to the formation of new tissues in the case of growing animals, while another part is used for the elaboration of special substances such as enzymes, hormones, pigments and so on. In the invertebrates some amino-acids are excreted unchanged, but the loss of appreciable amounts of unchanged amino-acids in the excreta of higher animals only takes place in rare and special disorders such as cystinuria.

Of the ammonia set free by deamination, part or all may be excreted unchanged. In most aquatic animals it is in fact the main nitrogenous constituent of the excreta, but in many fishes it is converted more or less completely into urea or, perhaps, into trimethylamine oxide. In higher vertebrate forms it is converted into urea or uric acid, and our next problem is that of finding out how these various substances are built up. Some trimethylamine oxide may perhaps originate in betaines present in the food, but in cases where it accounts for as much as a third of the total nitrogen excreted it must probably be largely synthetic in origin. Nothing is known for certain about its mode of synthesis, however, so that for all practical purposes we have only to consider ureotelic and uricotelic animals here.

As long ago as 1914 it was pointed out by Clementi

that the enzyme *arginase is present in the livers of ureotelic animals but absent where metabolism is uricotelic*. This suggested that there must be a close relationship between arginase and the synthesis of urea, but it was not until 1932 that Krebs and Henseleit were able to show just how intimate this connection really is. They showed that urea is built up by a cyclical mechanism which has come to be generally known as the 'ornithine cycle' (shown diagrammatically in Fig. 7) in which arginase plays a

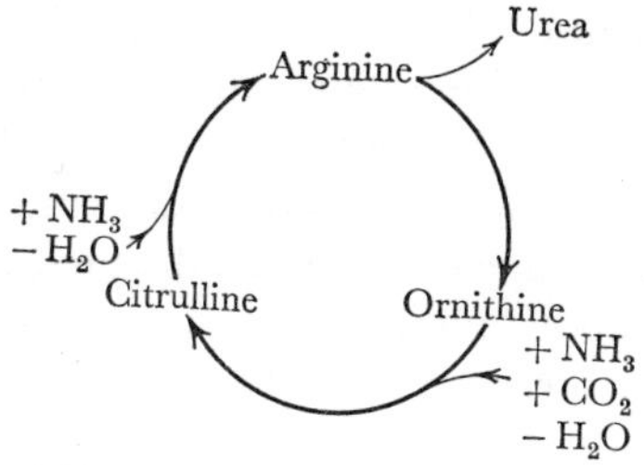

Fig. 7. The ornithine cycle. Urea is synthesised from ammonia and carbon dioxide, ornithine acting as a catalyst of the 'carrier' type. *Arginase* splits arginine into ornithine and *urea*.

part of central importance. Krebs and his collaborators worked first on mammals but later on other vertebrates, and found that the ornithine cycle is present in the livers of mammals, turtles and Amphibia, which are ureotelic and contain arginase, but absent from the livers of birds and most reptiles, which are uricotelic and contain no arginase. Thus ureotelic metabolism, liver arginase and the ornithine cycle go hand in hand.

The teleosts, though not ureotelic, contain arginase but not the ornithine cycle. The rest of this important mechanism must presumably have been added later in the process of evolution as an adaptation to limitation of water supply which, as we saw, called forth the production of urea in the Amphibia. But the elasmobranch fishes are particularly interesting in this connection, for in them urea plays a very large and important part (see p. 40). It is accordingly very interesting indeed to find that arginase is not here confined to the liver, but is present in practically every tissue of the body. Recent work has shown that the ornithine cycle almost certainly operates in elasmobranch liver.

There is some circumstantial evidence that tissues other than the liver may also be involved in these fishes. It is well known that if the liver of a dog, say, is extirpated, production of urea ceases, but if the same operation is carried out on an elasmobranch, the production of urea continues without interruption. This must presumably mean that urea can be synthesised all over the body and not only in the liver, suggesting that the ornithine cycle like arginase itself, is present practically everywhere in the body. We are therefore justified in concluding that the synthesis of urea in ureotelic vertebrates is in all cases carried out by the ornithine cycle, and that this mechanism is, as a rule, confined to the liver.

Although the livers of uricotelic organisms contain no arginase, small amounts of that enzyme are present in the kidney. Arginase, of course, splits arginine into urea and ornithine, and although urea itself is useless to uricotelic animals, the birds and perhaps the reptiles too make good use of the ornithine so produced. Benzoic acid occurs frequently in their food and is poisonous. It will be remembered that the mammals detoxicate benzoic acid by converting it into the non-poisonous hippuric acid; the birds, however, conjugate it with ornithine to give ornithuric acid (dibenzoylornithine).

We know a great deal about the mechanisms whereby uric acid is synthesised. This synthesis is a complex process involving about 15 stepwise reactions and is too complicated to be considered in these pages. One thing is certain, however; that urea is not, as was once believed, an intermediary in the process. Indeed it is difficult to see how, if urea were really an intermediate, it could be formed in the liver of the bird, which contains no arginase and from which the ornithine cycle is totally absent, unless some entirely different mechanism were present.

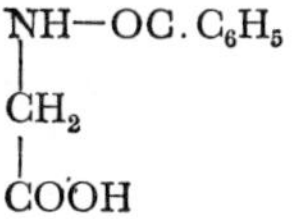

Hippuric acid

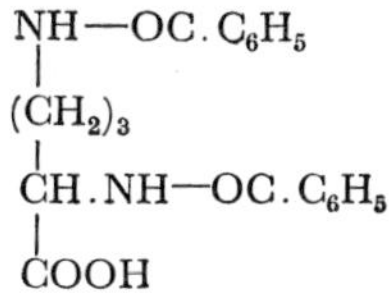

Ornithuric acid

A partial answer to the question of how uric acid is synthesised is worthy of note, however. It has been known for some years that the liver of the pigeon can use ammonia and make from it a substance which the kidney then turns into uric acid. Isolation of the precursor showed that it is hypoxanthine. In most birds the enzyme xanthine oxidase is present in both liver and kidney and, under its influence, hypoxanthine is oxidised to uric acid.

With regard to the invertebrates it is known, as we have pointed out, that certain terrestrial groups synthesise uric acid. As yet we know very little about the biochemical processes involved in these cases, but it is only a matter of time before this, like many other processes, will be analysed and understood.

4. *The metabolism of the purines*

Apart from α-amino-N, the chief sources of nitrogen are the nucleic acids. On hydrolysis of nucleic acid, bases of the purine and pyrimidine groups are obtained, but although a good deal is now known about the metabolism of the pyrimidines we shall confine our attention to that of the purines. These bases are set free by the digestion of nucleic acids, and are known as adenine and guanine respectively. Both are amino compounds and undergo deamination at the hands of specific enzymes known as adenase and guanase to give hypoxanthine and xanthine, and these are then oxidised to uric acid under the influence of a third enzyme, xanthine oxidase. The complete scheme for the conversion to uric acid may be represented as follows:

Adenine (6-aminopurine)		Guanine (2-amino-6-oxypurine)		
↓		↓		
Hypoxanthine (6-oxypurine)	→	Xanthine (2-6-oxypurine)	→	Uric acid (2-6-8-oxypurine)

Not all animals possess all three of the enzymes necessary to carry out the complete series of changes, so that free adenine or guanine may appear in the excreta. Thus pigs sometimes suffer from a disease known as 'guanine gout', which is due to the fact that guanase is absent. Guanine is therefore not metabolised and may accumulate and cause symptoms similar to those caused by uric acid in gout proper. As a general rule it may however be said that the higher animals are capable of converting ingested purines into uric acid. The uric acid so formed is excreted as such by man, by the higher apes, and also, though not in its entirety, by the Dalmatian dog. The latter does, however, have urico-oxidase but has a 'leaky kidney' as far as uric acid is concerned. Uric acid leaks out into the urine before the urico-oxidase of the liver has had time to oxidise it. But this is the exception rather than the rule, for the rest of the mammals possess urico-oxidase, which breaks down uric acid to give allantoin, and this substance is the usual end-product of purine metabolism among mammals. Why urico-oxidase should be missing from the tissues of the higher apes, and why there should be what appears superficially to be a deficiency of this enzyme in the tissues of the Dalmatian dog, is a mystery; if only man had not in some way lost his urico-oxidase, gout would be an unknown disorder. The possession of normal urico-oxidase activity is evidently an inheritable dominant factor among dogs, for puppies got by crossing a Dalmatian with a dog of another breed are found to be capable of breaking down uric acid and converting it quantitatively instead of only partially into allantoin.

Urico-oxidase is absent from uricotelic animals. This is almost to be expected, for it seems unlikely *a priori* that an animal which synthesises uric acid as its main nitrogenous end-product would possess an enzyme capable of undoing what has already been done. This important rule was first pointed out by Przylecki, and we may combine it with the other important rule due to Clementi (p. 67) in the statement that *neither arginase nor urico-oxidase occurs in the liver of uricotelic animals.*

Lower down among the vertebrates we again find urico-oxidase at work, and this time its action is succeeded by that of two more enzymes—allantoinase, which turns allantoin into allantoic acid, and allantoicase, which turns allantoic acid into urea and glyoxylic acid. The complete scheme of uricolysis is thus as follows:

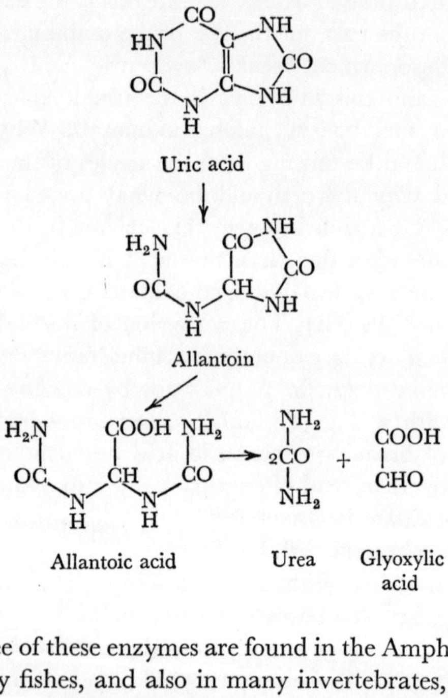

All three of these enzymes are found in the Amphibia and in many fishes, and also in many invertebrates, notably among the Echinodermata; in such cases the end-product of purine metabolism is urea. It might be mentioned that the enzyme urease has been found in a number of

Table 6

	End-product of						
	Protein metabolism	Purine metabolism	Liver arginase	Ornithine cycle	Xanthine oxidase	Urico-oxidase	Allantoinase and allantoicase
Mammalia	Urea	Allantoin[1]	+	+	+	+[2]	–
Aves	Uric acid	Uric acid	–	–	+	–	–
Reptilia:							
Snakes, lizards	Uric acid	Uric acid	–	–	+	–	–
Turtles	Urea ± uric acid	Allantoin?	+[3]	+[3]	+	?	?
Amphibia	Urea[4]	Urea	+	+	+	+	+
Pisces:							
Elasmobranchii	Urea	Urea	+	+	+	+	+
Teleostei	Ammonia	Urea	+	–	+	+	+

[1] Uric acid in man, higher apes and, though only in part, Dalmatian dog.
[2] Absent from man and higher apes (see p. 71).
[3] In some species.
[4] See also Table 5.

Invertebrata, notably in the snail family, but its function in the animal kingdom is still obscure.

The most important facts dealt with in this chapter have been summarised in Table 6 and Fig. 8, and a careful examination of these will serve to integrate our knowledge of the main lines of nitrogenous metabolism as they occur in members of the animal kingdom.

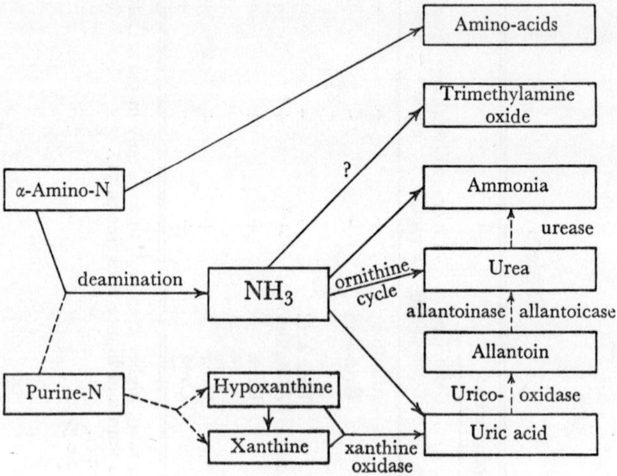

Fig. 8. Nitrogenous metabolism in animal organisms. The minor processes are denoted by broken lines.

CHAPTER V

1. *The distribution of nitrogenous bases*

MANY different nitrogenous bases of one kind and another have been isolated from the tissues of various animals. In some cases we find that a given compound is fairly sharply confined to certain groups of animals, and can draw conclusions as to the probable relationships between the animals constituting these groups or about the function of the particular compound. We have already seen that large amounts of urea are to be found in the blood and tissues of the elasmobranch fishes, a fact which separates them quite sharply from the other vertebrate groups with which we have to deal here. Similarly, we have seen that trimethylamine oxide is found in marine but never in fresh-water fishes, and this has enabled us to discuss the possible significance of the compound in relation to our knowledge of its distribution.

Now trimethylamine oxide is not entirely confined to the fishes. It is said to form an important constituent of the urine of cephalopods, while as much as 200 mg. per cent has been found in the muscles of Crustacea such as the lobster, *Homarus*. The characteristic unpleasant odour of dead fish, which is not detectable in fresh-water species, is largely due to the liberation of trimethylamine itself from its oxide, probably by bacterial action, shortly after the death of the animal.

A series of compounds related to trimethylamine oxide have been isolated from various invertebrates from time to time, and the formulae of some of them are given below. The principal animals in which their occurrence has been demonstrated are also indicated, but we cannot at present do more than guess at their function. Several different betaines have been obtained, some of which,

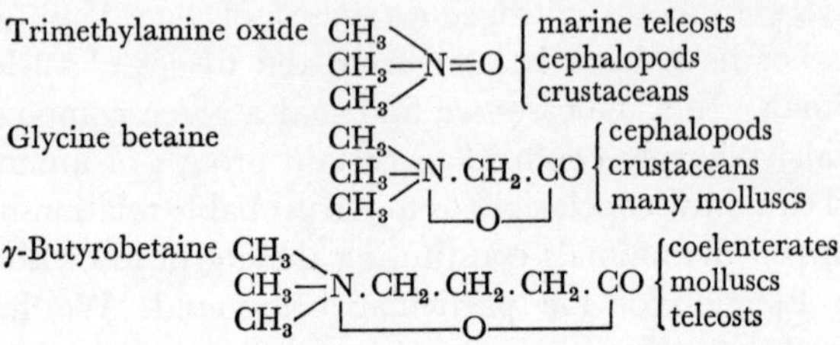

e.g. stachydrin and trigonelline, contain ring structures. The presence of trigonelline, which is a commonly occurring plant alkaloid, has been demonstrated in

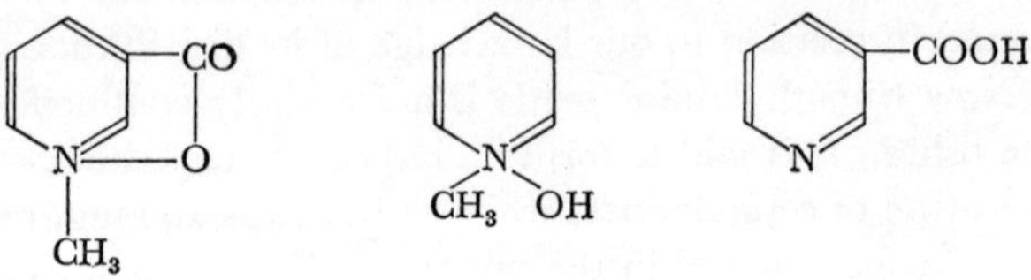

Trigonelline N-methylpyridine Nicotinic acid

several animals while the closely related N-methylpyridine is widely distributed in the animal kingdom, and these two compounds are especially interesting on account of their relationship to nicotinic acid, a substance which came into a prominent position with the discovery that its amide is a constituent of cozymase. Later researches showed that the coenzymes of lactate and many other dehydrogenases and the cozymase of yeast are identical and it is certain, therefore, that nicotinic acid derivatives are of great importance and that their distribution is very wide indeed.

Another compound of particular interest is the tetramethylammonium hydroxide obtained from a sea anemone. It acts like curare and was once thought to be responsible for the paralysing 'sting' which many of these animals are capable of inflicting. Two other compounds which occur very widely are taurine and glycine. Taurine is derived in mammals from cysteine by oxidation at the —SH group and decarboxylation, and is perhaps an end-product of sulphur metabolism. It has been found in all the groups of the Mollusca, in numerous Echinodermata, in a number of annelid worms and, in small quantities, even in vertebrate muscle. Its function is unknown.

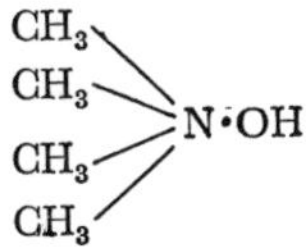

Tetramethylammonium hydroxide

It is frequently found that, although the osmotic pressure of the blood of invertebrates is the same as that of the surrounding sea water, the salt content of the blood is a little less, indicating that a part of the total osmotic pressure of the blood is due to non-electrolytes. It is thought that the non-electrolytes in question may be substances like taurine and the betaines, and in some cases the quantities of these substances actually found are about enough to account for the observed results.

Most interesting of all, however, are the nitrogenous

bases derived from guanidine, some of which, together with their formulae, are mentioned in the following list:

Guanidine	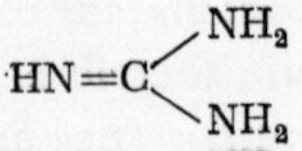	(Cephalopods)
Methylguanidine	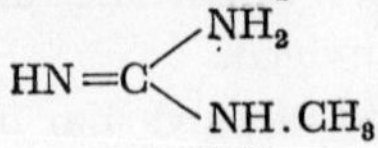	(Teleosts)
Glycocyamine	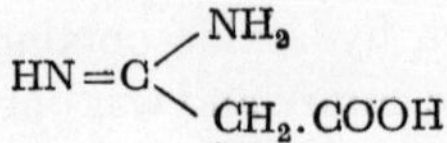	(Some annelids)
Creatine	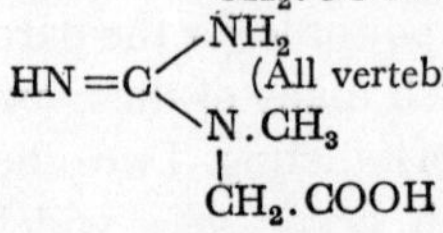	(All vertebrates; some annelids)
Arginine	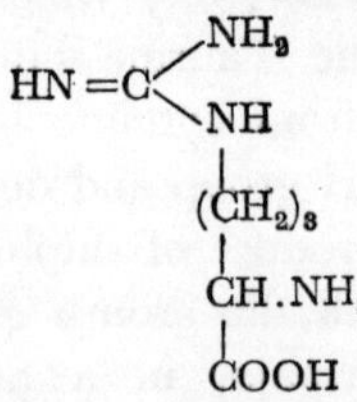	(Many invertebrates)
Agmatine	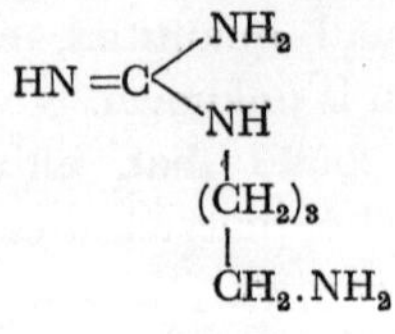	(Sponges)

Other guanidine derivatives too are known, but we shall confine ourselves here to the two best known members of the group, arginine and creatine. The functions of the others are still largely unknown. It can safely be said that creatine is present in the muscles of all classes of vertebrates, but that satisfactory evidence of its occurrence among invertebrates was, until fairly recently, practically non-existent. On the other hand, the muscles of invertebrates usually contain large amounts of arginine, only traces of which are to be found in the tissues of vertebrates except, of course, in the combined form in which it occurs in proteins. The distribution of these two compounds has been extensively studied, and the conclusion reached by Hunter when he reviewed the evidence in 1928 was that creatine is confined to members of the Vertebrata and arginine to the Invertebrata, a generalisation of the utmost comparative interest. Why arginine should thus be replaced by creatine as we pass from invertebrates to vertebrates we do not know for certain. It will be remembered that the enzyme arginase came into great prominence with the development of ureotelism among vertebrates, and perhaps arginine could no longer be used with safety on account of the danger that it might be hydrolysed by the enzyme, and possibly this is the reason. It was at one time believed that arginase does not occur among invertebrates, a view which fits well with that just suggested, but it is now known that the enzyme does occur frequently among these animals although, as a rule, its activity is small.

2. *The distribution of phosphagens*

In the same year that saw the publication of Hunter's book, the function of creatine was discovered. It was shown to be present in the muscles in the form of a phosphoric acid complex which received the name of phosphagen, and to play a part of great importance in the chemical events associated with muscular contraction.

These results have since been abundantly confirmed. A little later it was found that this compound, phosphocreatine, appeared to be exclusively confined to the vertebrates, its place in invertebrate muscle being taken by an arginine compound of similar structure and physiological importance. The formulae of these two important compounds are the following:

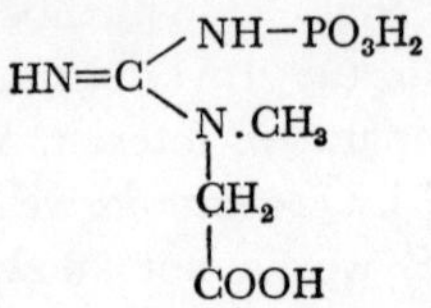

Phosphocreatine

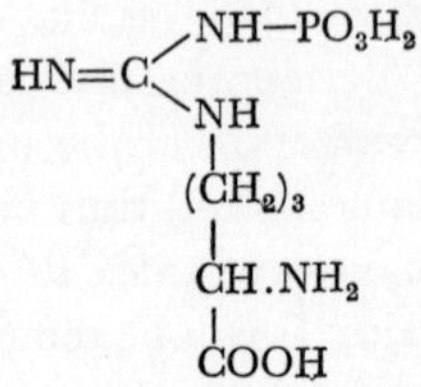

Phosphoarginine

This is not the place to go into the nature of the reactions into which these compounds enter, but it must be pointed out that they behave in exactly the same way and play the same part in muscular contraction. The question of their distribution is what particularly concerns us here, and we may say at once that, with certain important exceptions, *phosphocreatine only is found in the muscles of vertebrates while only phosphoarginine is found in the muscles of most invertebrates.*

In what follows we shall regard the phosphagen of invertebrates as being the arginine compound, and refer to it as AP for the sake of brevity. The creatine derivative will be referred to as CP.

The sudden replacement of AP by CP as we pass from invertebrates to vertebrates is a matter of very great interest, since, by finding out the precise point at which the change-over takes place, we might hope to be able to throw fresh light on the much discussed problem of the ancestry of the vertebrates. There exists a group of curious animals of particular interest from this point of view, namely the Protochordata, which are regarded by morphologists as in a sense on the border-line between the vertebrates and the invertebrates. None of them possesses the true backbone which is the main morphological characteristic of the vertebrates as a whole, but they all, at one time or another during their life, possess a primitive kind of backbone known as a notochord. Before going into the biochemical aspects of their importance it is desirable to consider their relative claims as possible ancestors of the vertebrates, and in order to get some idea of their appearance the reader should look at Fig. 9 in which the adult forms are depicted.

The Protochordata fall into three main groups, Tunicata, Cephalochorda and Enteropneusta. The Tunicata or sea squirts, as an example of which *Ascidia* may be taken, hatch from the egg as tadpole-like larvae, in the tail of which a well-developed notochord is present. On metamorphosis this notochord is completely lost and with it all signs of vertebrate affinity. Indeed, the sea squirts were at one time classified with the Mollusca. In the Cephalochorda, of which the lancelet, *Amphioxus*, may be taken as an example, there is a notochord which runs the whole length of the body and is present throughout the life of the animal, so that on the whole *Amphioxus* seems to be more closely related to the vertebrates than to any known invertebrate type. The Enteropneusta, represented by *Balanoglossus*, show strong resemblances both to

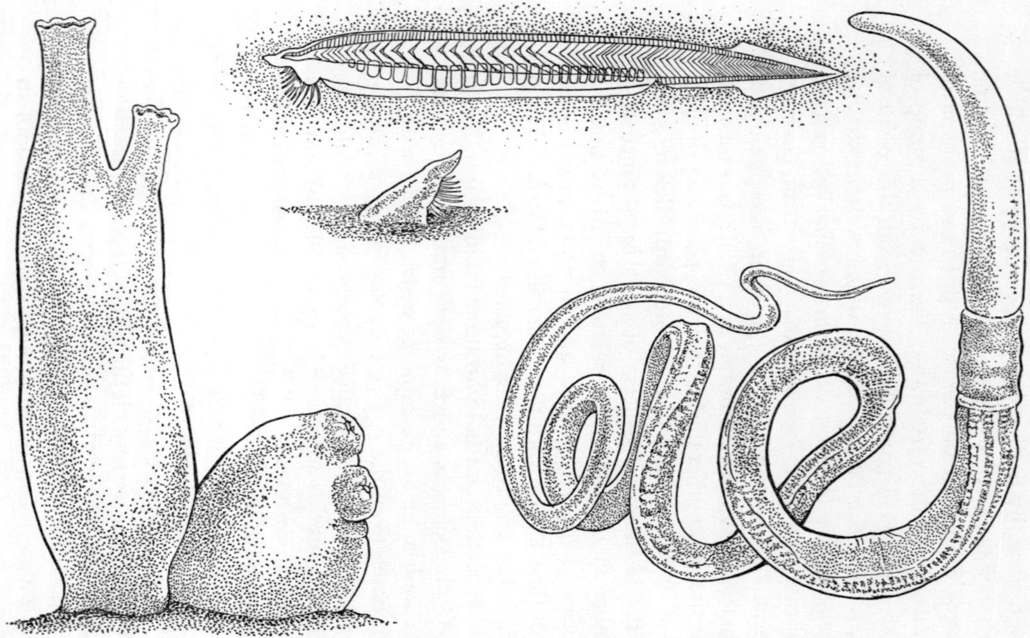

Fig. 9. Protochordates: on the left two Tunicates, on the right an Enteropneust, *Balanoglossus*, and above two specimens of *Amphioxus* (one nearly buried in the sand)—redrawn from Wells, Huxley and Wells.

(From *The Science of Life*, by permission of the authors and Messrs Cassel & Company.)

the vertebrates and to the invertebrates, and it is this group which the great biologist, William Bateson, regarded as the most likely of the Protochordata to represent a link between the vertebrates on the one hand and the invertebrates on the other. In its larval form *Balanoglossus* so closely resembles the larvae of the Echinodermata that it was actually classified with the echinoderm larvae before the adult form was discovered. The adult possesses a short but well-defined notochord in the region of the 'collar', and has other structural features which show that it is closely related to the vertebrates. On these grounds it is generally held to-day that the Enteropneusta represent a link between the Echinodermata on the one hand and the Vertebrata on the other.

We may now pass on to see what evidence comparative biochemistry has to offer towards the solution of the problem of the ancestry of the vertebrates. A large number of invertebrates has now been examined and the nature of their phosphagen determined. Phosphocreatine is, as we have pointed out, practically confined to members of the Vertebrata, but there are two noteworthy exceptions. As Table 7 shows, of all the many invertebrates and 'border-line' animals studied, two groups have long been known to contain both phosphoarginine and phosphocreatine and these were precisely the two which would be expected to ally themselves most closely with the vertebrates, namely the Echinodermata and Enteropneusta. The classical morphological evidence thus received support from a rather unexpected source, the work of the comparative biochemist. It was interesting indeed to find that perhaps we owe more to the prickly sea urchins and to the worm-like *Balanoglossus* than to any of the crabs, lobsters, worms and other invertebrates which inhabit the sea to-day. This makes a pretty story. But some twenty years later it turned out that comparative biochemistry had not yet been comparative enough at that time.

Table 7. *Distribution of phosphagen in the animal kingdom*

	AP	CP
Platyhelminthes	+	−
[1]Annelida	−	(+)
Arthropoda	+	−
Mollusca:		
Lamellibranchiata	+	−
Cephalopoda	+	−
Echinodermata:		
Crinoidea	+	−
Asteroidea	+	−
Holothuroidea	+	−
Echinoidea	+	+
Ophiuroidea	−	+
Protochordata:		
Tunicata	−	+
Enteropneusta	+	+
Cephalochorda	−	+
Vertebrata: all classes	−	+

A new and concentrated attack upon the annelid worms showed that, contrary to expectation, none of them contain phosphoarginine. Instead phosphoglycocyamine, phosphocreatine and phosphotaurocyamine were present—an object lesson warning us not to try to deduce too much from too few species!

The annelids seem to be a remarkably versatile group

[1] The presence of other hitherto unknown phosphagens has been demonstrated in marine annelids.

for, apart from those already listed and which occur in marine species, the (terrestrial) earthworms possess a peculiar guanidine derivative that seems to be entirely their own. Most recently yet another new compound has been discovered in leeches. It may well be that still other hitherto unknown phosphagens may occur here or there in the animal kingdom.

Before we leave the question of the distribution of phosphagen, it is interesting to see to what extent phosphagen is connected with forms of activity other than that of muscle. Small quantities of phosphagen have been discovered in nerve, but very little is known of the part it plays in the conduction of the nervous impulse, if, indeed, it plays any at all. But phosphagen is known to be present in the electric organs of the electric ray, *Torpedo*, a fish which is capable of delivering severe electric shocks of the order of 300 volts on open circuit. The cells of this organ arise from pre-muscular tissue in the embryo and they resemble the cells of true muscle in many respects. Thus they act on the all-or-none principle, can be tetanised, and are similarly affected by various drugs. Whereas muscle is specialised for the performance of mechanical work, the importance of electrical changes being negligible, the situation is exactly reversed in the case of the electric organ, and it is extremely interesting to find that these two effector organs make use of very similar and probably identical chemical mechanisms in spite of the great difference in the nature of their respective responses to stimulation. Phosphagen is also present in the electric organs of another ray, *Raia clavata*, in that of the electric eel, *Electrophorus*, and in the electric tissue of the catfish, *Malapterurus electricus*.

At one time it seemed possible that phosphagen might play a part in ciliary activity, for it was found in several organisms in which locomotion is effected mainly or entirely by ciliary action. But a study of several kinds of Protozoa, the locomotion of which is due to cilia and flagellae alone, failed to reveal the presence even of traces of phosphagen and it now seems very improbable that phosphagen is concerned in this form of activity. Its presence in the developing embryos of the echinoid, *Paracentrotus*, for example, must probably be attributed to the presence of pre-muscular tissues from which the true muscle later develops rather than to the cilia which provide their sole mechanism for locomotion. The results obtained in the case of the Coelenterata have always been rather doubtful, which is unfortunate since, although they possess no mesoderm, which is the layer from which muscle ordinarily develops, these animals do possess muscular cells derived from other sources, and myoepithelial cells which discharge comparable functions. So far as the results go they appear to indicate that traces of a phosphagen of the arginine type may be present and arginine itself has been isolated from one or two species. Further investigations in this direction would be amply repaid.

The results so far obtained may be summed up in a few words. They show that wherever we find true muscle, one phosphagen or another is present too, and there is evidence from other sources to show that, in spite of the great specialisation which muscle has undergone for one purpose and another, the chemical mechanisms involved in supplying the energy for contraction have remained fundamentally the same. Phosphagen is present in smooth, cardiac and striated muscle throughout the animal kingdom. It is present also in the electric organs of certain fishes, and, so far as the evidence goes, always discharges the same function. More recent work has shown that the dynamic biochemistry of the muscles of the crab, octopus, sea-urchin and sea-cucumber, and also the electric organs of *Torpedo* and other electric fishes, resembles that of vertebrate muscle very closely indeed.

CHAPTER VI

1. *Respiration*

We have already devoted some space to a consideration of the blood as a medium upon the constant properties of which the cells of the living organism depend for the maintenance of working conditions. Nothing has been said of the blood as a medium which carries foodstuffs to and waste products away from the tissues, but in this chapter we have to deal with the carriage of the primary foodstuff, oxygen, and the removal of the primary waste product, carbon dioxide.

No one would deny that, with very few exceptions, all animal cells respire. That is to say, they take up oxygen and produce carbon dioxide. The latter is an acidic substance which, if allowed to accumulate in the tissues, would sooner or later cause the pH to fall to a lethal level, and an animal is therefore just as liable to choke through being unable to dispose of carbon dioxide as it is to suffocate from lack of oxygen. An animal must be able to do two things: first to secure an adequate supply of oxygen from its surroundings, and secondly to rid itself of carbon dioxide. In small, simple animals such as the platyhelminth worms, most of which are thin, flat animals less than half an inch long, the energy requirements are small, and the correspondingly small exchanges of oxygen and carbon dioxide can be carried out across the body surface, which is very large as compared with the bulk of the animal in such cases. But diffusion is a very slow process, and in larger animals, not only is the ratio of area to volume very much smaller, but the surface is too far away from the internal structures for oxygen to be able to pass in rapidly enough by simple diffusion. In practically all animals from the Annelida upwards we find a system of blood vessels and a heart of some kind to drive the blood round the body. Oxygen is obtained by having special respiratory organs in which the blood is brought into very close proximity to the external medium. Here oxygen diffuses rapidly into the blood and is carried round to the deeper tissues. In some animals, e.g. the frog and the earthworm, the skin itself can act as a respiratory organ, but as a rule we find specialised organs such as the gills of water-breathing animals and the lungs of air breathers. For purposes of argument we can divide the animal into three parts, the respiratory organ or organs, the blood, and the tissues. As a rule the blood circulates continuously through the body, but in a few animals—notably certain of the annelid worms—there is in the capillaries no 'throughway' circulation but only an ebb and flow of the blood.

Considerable modifications in the respiratory organs became necessary before animals could leave an aquatic in favour of a terrestrial environment. The advantages to be gained thereby included the exchange of a relatively poor oxygen supply for a very rich one. But respiration has always remained essentially aquatic. The respiratory epithelium, whether it be that of a gill or a lung, is always covered by a stationary aqueous layer. In the case of an aquatic organism this stationary layer is no more than a few molecules thick and offers little obstruction to the diffusion of oxygen. The epithelium of a lung, however, is protected from drying up by a mucous-aqueous film which is relatively thick in comparison. The limiting factor in the rate of passage of oxygen through the respiratory epithelium is the rate of diffusion, and the relatively greater thickness of the stationary layer through which oxygen must diffuse in the case of an air-breathing animal must mean that, other things being equal, aerial respiration is less efficient than the aquatic type. This relative inefficiency of the respiratory organ appears to have been compensated for by an increase in the area of the respiratory epithelium per unit body weight in air-breathing animals.

There exist several series of closely related animals, individual members of which show progressive specialisation of the respiratory organs for aerial respiration. Thus the four species of periwinkle, *Littorina*, inhabit successively higher levels in the marine littoral regions, and correlated with the corresponding necessity for air breathing, we find increasing vascularisation of the mantle cavity as we pass from *L. littorea* and *L. obtusata*, through *L. rudis* to *L. neritoides*. The highest degree of vascularisation is found in the terrestrial snails (e.g. *Helix*) and slugs (e.g. *Arion*), which are exclusively air breathers and have a mantle cavity with walls no more than $2\,\mu$ in thickness and very richly supplied with blood vessels. The hermit crabs of Tortugas afford another interesting series. Here the gills are found to have degenerated as their respiratory function has been taken over by vascularised areas in other parts of the body. Two species which live always below the low-water mark have 26 gills each, while a third species living at high-water mark has only 18. In a fourth species, which lives always above the high-water mark, the number of gills has been reduced to only 14.

The main question which we have to consider here is the efficiency of the blood in carrying oxygen from the respiratory organ to the tissues. Much valuable work has also been done on the other aspect of the respiratory function of the blood, namely, its capacity for carrying carbon dioxide away from the tissues.

In a very general kind of way we may say that, to be efficient, the blood must carry (*a*) enough oxygen to the tissues which require it, and (*b*) enough carbon dioxide away from the tissues in which it is produced. Condition (*b*) is complicated by the fact that, being acidic, carbon dioxide must be buffered during its passage through the blood in order to avoid dangerous changes in pH.

As we have seen in an earlier section, primitive blood probably consisted of little but plain sea water, and this could probably carry enough dissolved oxygen to meet the requirements of fairly simple animals. Later on, however, when animals became more active and greater supplies of oxygen were therefore called for, the oxygen capacity of the blood had to be increased in some way. The mammals are very active, warm-blooded creatures, and they are able to carry about 25 ml. of oxygen per 100 ml. of blood as compared with the 0·5 ml. or thereabouts which sea water could carry under similar conditions. Here, of course, the large oxygen capacity has been attained by the inclusion of haemoglobin in the blood, but several other substances have been successfully used by other animals for the same purpose, and we shall discuss these in a later section. For the present we must confine our attention to some of the properties which the blood must possess if it is to function efficiently. Not all animals live under the same conditions; oxygen is available in much larger quantities and, what is more important, at higher partial pressures, to animals which breathe air than to those which breathe water, and it is found that the properties of the blood differ accordingly. The major differences are due to the presence of different oxygen carriers belonging to the group of substances usually referred to as 'respiratory pigments' on account of the fact that all those so far discovered are coloured, but there is no obvious reason why colourless substances should not work equally well. Indeed, it is conceivable that respiratory 'pigments' exist which have escaped detection simply because they are not coloured!

Mammalian blood such as our own is an exceedingly complicated system, containing as it does astronomical numbers of red blood corpuscles and smaller numbers of white cells of various kinds. Several dozen proteins also are present in solution, together with a rich assortment of smaller molecules such as glucose, urea, as well as simple salts and ions. We can, however, give some account of the manner in which the respiratory function of a blood as complicated even as this originated and evolved and we have already seen that sea water was the probable starting-point.

What then are the principal properties of an efficient blood? In the first place it must react *reversibly* with oxygen. The blood must take up oxygen at the respiratory organs, where the partial pressure of oxygen is high, and give it up again equally readily to the tissues, in which the partial pressure is low. It follows, therefore, that a substance which acts as an oxygen carrier in the blood must be one that reacts reversibly with oxygen; in other words, the compound of the respiratory pigment with oxygen must be such that it readily dissociates. Pyrogallol, for example, combines with oxygen very readily indeed, but would clearly be useless as a respiratory pigment since it refuses to give it up again.

Most of the pigments so far studied give dissociation curves of the type shown in Fig. 10 when allowed to react with oxygen under conditions as nearly as possible identical with those obtaining in the blood of the animal itself. All the curves shown in Fig. 10 belong to a family which can be mathematically expressed by Hill's equation,

$$y = 100 \frac{kx^n}{1 + kx^n},$$

where y is the percentage of the blood present in combination with oxygen when the partial pressure or tension of the latter is x; k and n are constants, k measuring the affinity of the blood for oxygen, while the shape of the curve depends upon the value of n. If n is unity, the curve is hyperbolic like curve 1 in the figure, but as n increases the curve becomes more and more S-shaped, like curves 2, 3 and 4, and the shape of the curve is a matter of considerable importance. Consider the case of an animal, the blood of which takes up oxygen from a medium in which the oxygen tension is, say, T, and gives it up again to the tissues in which the tension is t. It will be seen from Fig. 10 that a blood with a hyperbolic dissociation curve (1) gives up but little of its oxygen when the oxygen tensions falls from T to t, while bloods 2, 3 and 4 give up progressively more and more for the same drop in oxygen tension and are accordingly more efficient. The more S-shaped the curve, the more efficient is the blood, and the value of n in Hill's equation therefore gives us a measure of the efficiency of the blood in this respect.

A blood will also be more or less efficient according as it is able to carry more or less oxygen. Table 8 gives a list of the approximate oxygen capacities of the bloods of various animal groups, and from these we can get some

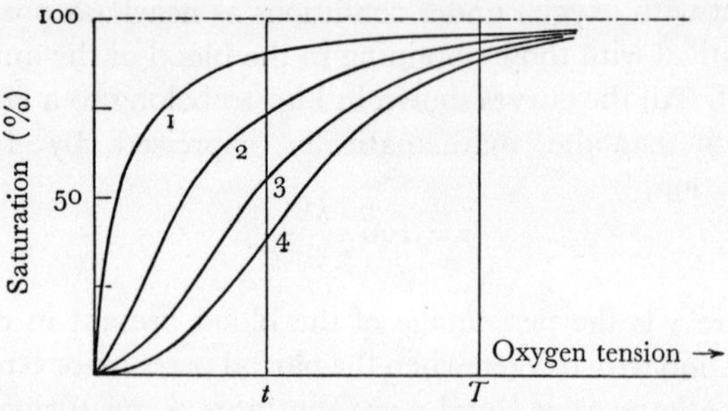

Fig. 10. A group of typical dissociation curves, to illustrate the importance of the constant n.

idea of the relative merits of each of the four respiratory pigments represented. Of the four, haemoglobin has been the most successful, enabling mammalian blood to carry as much as a quarter of its own volume of oxygen, while the runners up, chlorocruorin and haemocyanin, are far behind with the 9 and 8 volumes per cent respectively which they enable the bloods of certain annelids and cephalopods to carry. Mammalian haemoglobin is far and away the most successful of the respiratory pigments from this point of view, and Barcroft has written of it that 'but for its existence, man might never have achieved any activity which the lobster does not possess'.

The fourth property which we have to consider concerns the oxygen tension required to saturate the blood. All the dissociation curves so far obtained are asymptotic,

Table 8. *Oxygen capacities of some different bloods*

Pigment	Colour	Site	Animal	ml. Oxygen per 100 ml. blood
Haemoglobin	Red	Corpuscles	Mammals	25
			Birds	18·5
			Reptiles	9
			Amphibians	12
			Fishes	9
		Plasma	Annelids	6·5
			Molluscs	1·5
Haemocyanin	Blue	Plasma	Molluscs:	
			Gastropods	2
			Cephalopods	8
			Crustaceans	3
Chlorocruorin	Green	Plasma	Annelids	9
Haemerythrin	Red	Corpuscles	Annelids	2

so that theoretically the bloods are only fully saturated at an infinite pressure of oxygen, but we can bring the matter down to the level of practical affairs by considering the oxygen tension necessary to saturate the blood, not completely, but to the extent of 95 per cent. This we call the loading tension of the blood, usually represented by the symbol T_L. Now suppose that a given animal lives in a medium in which the oxygen tension available is, let us say, T_E; how must the values of T_L and T_E be related to each other? A glance at Fig. 11 shows that if T_E is less than T_L the blood will never be fully saturated in the respiratory organs, so that a part of its total oxygen capacity will be of no avail. If, on the other hand, T_E is greater than T_L, the blood will be fully saturated but the animal unable to take advantage of any of the external tension over and above that required to saturate the

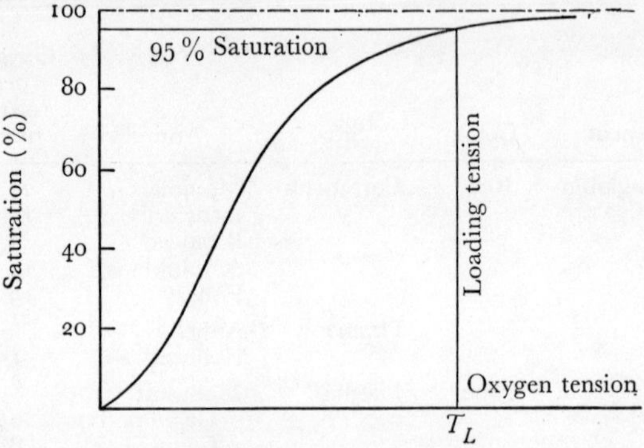

Fig. 11. Typical dissociation curve, to illustrate the importance of the loading tension.

blood; in other words a part of the oxygen tension available will be of no use. Any surplus pressure will only serve slightly to increase the amount of oxygen carried in physical solution in the blood. The greatest efficiency will therefore be attained by a blood of which the loading tension is about equal to the oxygen tension available in the environment.

The value of the loading tension is probably significant also in connection with the ease with which an animal can become adapted to aerial respiration. Migration into terrestrial surroundings has taken place frequently, and particularly so among the molluscs and the arthropods, both of which possess haemocyanin as their respiratory pigment. Bloods containing haemocyanin have, as a rule, a low loading tension as compared with those which contain haemoglobin. But we have seen that aerial respiration is an inefficient process as compared with aquatic respiration, and it would therefore seem that an animal going on to the land would find it better to have a blood with a rather low loading tension than one of which the loading tension was already high, for whereas a blood with a high loading tension might not become fully oxygenated during its passage through the respiratory organ, a blood whose loading tension was low would be reasonably certain to become fully saturated. Complete saturation of a blood with a high loading tension would require that the longer time taken for oxygen to diffuse through the stationary layer should be compensated for by allowing the blood to remain in close proximity to the source of oxygen for a longer period, and this would require enlargement of the respiratory surface, an adaptational process which in all probability could not be accomplished very quickly. Once established on the land and properly adapted to aerial respiration, a pigment such as haemoglobin would be more advantageous, since it would enable the blood to carry more oxygen and the animal to be more active than could otherwise be the case. Thus it seems likely that landward migration might perhaps be easier for an animal possessing haemocyanin than for one possessing haemoglobin but, migration once accomplished, haemoglobin would be the better of the two on account of the greater oxygen capacity which it can confer upon the blood.

Summing up, we are in a position to say that an efficient blood (*a*) reacts reversibly with oxygen, (*b*) has an S-shaped dissociation curve, (*c*) has the power to carry a large amount of oxygen per unit volume, and (*d*) has a loading tension which is about the same as the oxygen tension available in the environment.

All the known respiratory pigments fulfil the essential requirement of reacting reversibly with oxygen and, as a rule, the n value of the blood is greater than unity. The pigments enable the bloods of the animals in which they occur to carry more oxygen than could be transported in their absence, and the data of Table 8 exemplify this. In general, too, the loading tension is equal to or less than the oxygen tension available to the animal, but cases are known in which, while the first three conditions are fulfilled, the T_L values are far below the normal T_E values of the environment. This means, of course, that the pigment present in such cases is ordinarily fully oxygenated and therefore cannot play any part in the transport of oxygen. We shall discuss two such cases in more detail.

A certain aquatic snail, *Planorbis*, differs from other gastropods in having in its blood a haemoglobin instead of the usual haemocyanin, and the loading tension of this is only about 5–10 mm. *Planorbis* lives in slow-moving or stationary fresh water which is fairly well aerated as a rule, so that the haemoglobin must be present in the oxygenated condition. The animal gets all the oxygen it needs in physical solution in the blood. But if, as frequently happens in fresh waters of this kind, the oxygen tension should fall off for some reason, it would become increasingly difficult for the animal to obtain all the oxygen it needs, since there is then a smaller head of pressure to drive the gas into solution. When the tension falls to about 10 mm., however, the haemoglobin comes into action and the animal now derives its supplies of oxygen through the agency of its haemoglobin, just as does any other animal using a respiratory pigment. The possession of a pigment in this case is not ordinarily necessary, for sufficient oxygen can usually be got in simple solution, but it enables the animal to survive at times when the oxygen content of the water is unusually low and to penetrate into regions from which it would otherwise be excluded on account of the shortage of oxygen.

The lugworm, *Arenicola*, possesses a similar haemoglobin, and this animal lives in a burrow in the sand just above low-water mark, so that it is covered by well-oxygenated sea water most of the time. During this period its haemoglobin is in the fully oxygenated condition and the oxygen required is carried in simple solution in the blood. When the tide falls, the animal shuts itself up in its burrow and waits for the water to return, and meanwhile is cut off from external sources of oxygen, apart from oxygen dissolved in the water that drains past the burrow. The oxygen dissolved in the blood is soon used up, so that the oxygen tension in the blood falls sharply soon after the burrow has been closed. When the tension has fallen to about 10 mm., the combined oxygen of the haemoglobin becomes available, and it is mainly on this that the animal lives until the tide returns. It has been shown that a specimen of *Arenicola* contains about enough haemoglobin to carry it over the periods during which it is cut off from the dissolved oxygen of the sea water. In this case, therefore, we have to look upon the respiratory pigment as a device for storing up oxygen against the time when oxygen would not otherwise be available.

Respiratory pigments, then, have been used to meet the recurrent emergencies of everyday life as well as to keep up a constant and copious supply of oxygen to animals whose lives are not periodically endangered by a shortage of oxygen.

2. *Respiratory pigments*

Most of the naturally occurring respiratory pigments fall into one or another of four main groups known respectively as haemoglobins, haemocyanins, chlorocruorins and haemerythrins, and of these the first two are by far the most widely distributed. The individual members of each group are closely related to each other, but they are individually distinct from each other in crystalline form, loading tension, span, solubility and so on. Mouse haemoglobin, for example, is quite distinct from rabbit haemoglobin, but both of them, in common with all the other haemoglobins, can be split into two parts, an active or prosthetic group, and a protein. The prosthetic group of the haemoglobins is a complex iron compound called haem, while the protein part is known as globin. Haem is, as it were, the business end of the haemoglobin molecule, for it is here that the reaction with oxygen takes place.

All haemoglobins have the same haem, individual differences being due to differences in the nature of the globin. In the free state, haem is rather an insoluble compound which very readily reacts with oxygen but can only be separated from it again with considerable difficulty. Globin has the effect of making the prosthetic group more soluble and of making its reaction with oxygen a reversible one. It has been spoken of as 'the fine adjustment' of the haemoglobin molecule, for it is the globin which decides the loading tension and other properties of the complete molecule.

Table 9

	Haemoglobin	Chlorocruorin	Haemocyanin	Haemerythrin
Colour	Red	Green	Blue	Red
Metal	Fe	Fe	Cu	Fe
Prosthetic group	Haem	Haem	Polypeptide	?
Molecule oxygen per atom metal	1:1	1:1	1:2	1:3
General properties	Sharp spectral bands Forms CO compound	Sharp spectra bands Forms CO compound	No sharp spectral bands Forms CO compound	No sharp spectral bands No CO compound
Occurrence	Corpuscles or plasma	Plasma	Plasma	Corpuscles or plasma

The prosthetic group of the chlorocruorins is also a haem, but this is not the same as that of haemoglobin although built on the same general plan. Iron is present also in the prosthetic group of the haemerythrins, but apart from this our knowledge of this last group of pigments is very limited. The haemocyanins contain copper, the prosthetic group consisting in this case of an atom of copper in combination with a polypeptide believed to contain a molecule each of tyrosine and leucine, three molecules of serine, and a sulphur compound which has not yet been identified. These facts, together with some other points of importance, are summarised in Table 9.

It is interesting to notice that chlorocruorin and haemocyanin are only found in solution in the blood plasma, whereas haemoglobin and haemerythrin may occur either in solution or in corpuscles. Moreover it is a curious fact that while haemoglobin has a molecular weight of about 68,000 (determined by ultracentrifugation) when confined to corpuscles, pigments dissolved in the plasma have molecular weights of the order of 2,000,000. It should also be noticed that whereas a molecule of oxygen is carried by only one atom of iron in the haemoglobins and chlorocruorins, three atoms of iron are required in the case of the haemerythrins and two atoms of copper in that of the haemocyanins. These facts probably go far towards explaining why it is that, on the whole, haemoglobin has proved to be the most successful carrier of oxygen. Chlorocruorin might conceivably have been more successful than haemoglobin in this respect, for it contains a higher percentage of iron than does haemoglobin, but for some reason it is very limited in its occurrence and is never found except in solution in the plasma.

The distribution of these pigments among members of the animal kingdom is a matter of great zoological interest. Chlorocruorin and haemerythrin are closely confined to certain zoological groups, the former appering only in a small group of polychaete worms (*Chlorhaemidae*) and the latter in certain gephyreans and a single polychaete. Haemocyanin is similarly confined to the Mollusca (Cephalopoda and Gastropoda) and to the Arthropoda (especially the Crustacea), but haemoglobin appears to be distributed in a very haphazard fashion without the slightest regard to zoological classification. It is present in all the vertebrates, in a few holothurians, several crustaceans and at least two insects, in several lamellibranchs and one gastropod (*Planorbis*), in many annelid worms (e.g. *Arenicola*), in several parasitic nematodes (e.g. *Ascaris*, the common round-worm) and even in two species of Platyhelminthes. For a long time it was difficult to understand how so many animals of such different kinds can produce haemoglobin, while the ability to produce the other pigments seems to be very sharply confined to particular groups of closely related animals. But it is now known that practically all living cells contain cytochrome, which comprises a group of haem pigments the haem of which is certainly very nearly related to that of haemoglobin. Since cells in general are capable of elaborating the haem of cytochrome, it is very probable that they can also produce the haem of haemoglobin if the need arises. This argument, reasonable though it may appear, is not entirely watertight, for the component known as cytochrome a_3 has as its prosthetic group a haem which is identical, apparently, with that of chlorocruorin. This being the case it seems curious that chlorocruorin should be so much rarer than haemoglobin.

In the vertebrates, haemoglobin acts as a respiratory pigment in the ordinary sense, but in the invertebrates it seems to have arisen independently and to have been retained in many different species as an adaptation which enables them to withstand longer or shorter periods of oxygen deficiency. In general, the haemoglobins of invertebrates are markedly unlike those of the vertebrates in several important properties such as molecular weight and loading tension. For this reason it has been suggested that they should be called erythrocruorins, the name haemoglobins being restricted to the compounds found in vertebrate bloods.

To these must be added a third group of haemoglobins, the so-called muscle haemoglobins, or *myoglobins*, which are responsible for the colour of red meats and we already know many interesting facts about them. Their most important known property is their possession of a loading tension much lower than that of the blood haemoglobins. In resting muscle the myoglobin is present in the oxygenated state, but rapidly becomes reduced when the muscle contracts and oxygen is required for oxidative processes within the cells. It thus acts as an instantaneously available store of oxygen which must be of great importance in the oxygen supply of the muscle, and in this connection it is interesting to notice that the amounts of myoglobin found in different muscles are roughly proportional to the activities of the muscles themselves. Owing to its lower loading tension, the reduced myoglobin can be again oxygenated at the expense of the more readily dissociable oxyhaemoglobin of the blood, so that a continuous supply of oxygen to the active muscle is kept up, while a store of immediately available oxygen is held in readiness in the resting muscle. Myoglobin thus acts as the second member of the chain of haem-containing pigments, blood haemoglobin, myoglobin and cytochrome, along which oxygen is carried from the external atmosphere to the innermost oxidative mechanisms of the muscle cell.

A pigment of considerable interest is found in the blood of certain Tunicata. In addition to a large number of colourless ones, smaller numbers of green, blue and orange corpuscles are found, and these are thought to contain different oxides of vanadium (V_2O_3, V_2O_4 and V_2O_5 respectively) in combination with protein. This so-called vanadium chromogen contains about 10 per cent of vanadium, and is thought to be a pyrrol compound of some kind. It takes up oxygen with great readiness and gives it up again freely in acid solution, and as the corpuscles contain about 3 per cent of free sulphuric acid, the compound might perhaps qualify as a respiratory pigment. Its action in this capacity has been seriously questioned however and certainly has not been satisfactorily demonstrated. The presence of vanadium in sea water cannot be directly demonstrated by chemical methods, yet the Tunicata contrive to accumulate considerable quantities of the element during the course of their lives. This is rather an astonishing performance and so, for that matter, is the retention of 3 per cent of sulphuric acid inside corpuscles suspended in a neutral plasma. These peculiarities, together with the fact that their mantle consists of nearly pure cellulose, and the ability of these animals to 'de-grow' and de-differentiate during bad seasons and develop once more when conditions are again favourable, combine to place the Tunicata among the most interesting of all the lower animals.

The accumulation by living organisms of elements of which only traces are present in the environment is a phenomenon which is by no means peculiar to the Tunicata. Sea water contains about 10 mg. Cu per cubic metre whereas haemocyanin-containing bloods may contain 2–20 mg. Cu per 100 ml., so that the concentration factor in this case is of the order of 10,000. Fox and Ramage carried out spectrographic analyses of a number of animal tissues and showed that certain elements frequently occur in unexpectedly large amounts, notably Fe, Cu, Mn, Ni, Co, Pb, Ag, Cd, Li, and Sr, all of which are present only in traces in normal biological environments.

In conclusion mention may be made of pinnaglobin, and sycotypin, obtained from the molluscs *Pinna* and *Sycotypus*. These products, which were at one time alleged to contain manganese and/or zinc and to act as respiratory proteins, are now known to have been merely extremely crude protein-containing materials, probably of no respiratory significance whatever.

3. *Respiratory catalysts*

All the pigments discussed in the preceding section are compounds which react reversibly with oxygen, and are concerned with the carriage of oxygen from the respiratory organs to the tissues. There are other pigments which have a respiratory function but which are usually referred to as respiratory catalysts; they are concerned as carriers of hydrogen in the respiratory processes of the cells and tissues rather than with the carriage of oxygen to them. The best-known member of this group is the group of haem pigments, collectively known as *cytochrome*, which are found in all aerobic cells and play a part of fundamental importance in their respiration. Other haem derivatives related to cytochrome are the enzymes catalase and peroxidase, and mention may also be made of actiniohaematin, a pigment found in certain coelenterates and thought at one time to have a respiratory function. Helicorubin, the bright red pigment found in the gut of the snail *Helix*, is very widely distributed among the Mollusca and occurs occasionally in other animals, but its function is still quite obscure. It is, however, related to cytochrome in some respects and is mentioned here only for that reason.

Cytochrome is closely related to haemoglobin and chlorocruorin, for the prosthetic group of reduced cytochrome *b* is perhaps identical with that of haemoglobin and that of a_3 with the haem of chlorocruorin. The haem of cytochrome *c*, the most abundant component, is closely related to that of haemoglobin though not, apparently, identical with it. The prosthetic group, haem, contains ferrous iron which can be oxidised to the ferric state. Thus the prosthetic groups of oxidised cytochrome *b* and oxidised haemoglobin (methaemoglobin) are identical with each other and with the well-known substance haematin. In the living cell, *oxidised cytochrome is reduced by suitably activated molecules of organic metabolites* such as succinate and lactate. *In the presence of the enzyme cytochrome oxidase the reduced cytochrome can react with oxygen and so return to the oxidised form*, so that each molecule of cytochrome can be used over and over again and, moreover, with extreme rapidity.

Since the re-oxidation of cytochrome requires the catalytic action of cytochrome oxidase, we ought, by inactivating this enzyme, to be able to find out to what extent cytochrome is involved in the respiration of any given tissue. It so happens that this enzyme is almost completely inhibited by *M*/100 cyanide, and cyanide at this concentration inhibits the respiration of most cells and tissues to the extent of 60–90 per cent. In such cases, therefore, cytochrome appears to be the main line of communication between hydrogen, coming from the substances undergoing intracellular oxidation, and oxygen brought to the cell from the environment. The mechanisms which bring about the reduction of cytochrome have been intensively studied over many years, and although this is not the place to discuss them in detail, a few indications of their general nature may not be out of place.

Substances undergoing intracellular oxidation are seldom, and perhaps never, oxidised by the addition to them of oxygen. *Biological oxidations proceed as a rule by the removal of hydrogen atoms from the molecules undergoing oxidation*, and these atoms are usually removed in pairs. The enzymes which activate their substrates in such a way as to make this removal possible are known as *dehydrogenases*, but no removal of hydrogen takes place, even from activated substrate molecules, unless there is also present another substance to which the hydrogen atoms can be transferred. Substances which can act in this capacity are spoken of as *hydrogen acceptors*. For experimental purposes it is possible to replace the natural hydrogen acceptors of the tissues by reversibly oxidisable dyestuffs such as methylene blue, cresyl blue and many more, but needless to say, none of these dyes occurs naturally in the tissues. Like most enzymes, the dehydrogenases are very specific, but in addition to their specificity towards the substrate they also show a high order of specificity with respect to the hydrogen acceptor. Thus only a few dehydrogenases (Type 1) are able to catalyse the transfer of hydrogen from their substrates to molecular oxygen, and such enzymes as do possess this power are spoken of as *oxidases*. As examples of this group we may mention D-amino-acid oxidase. This is a conjugated protein which possesses as prosthetic group a dinucleotide of adenine (a purine, see p. 70) and a yellow pigment called riboflavin, which is identical with a member of the B_2 vitamin complex.

The principal types of dehydrogenase systems may be summarised thus:

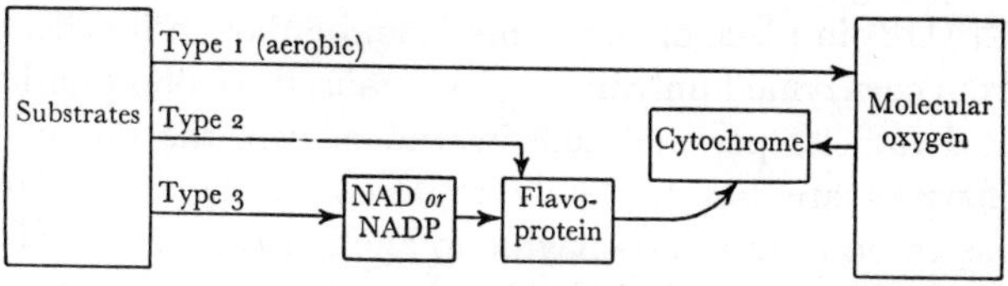

Arrows from left to right indicate transfer of hydrogen; from right to left they indicate transfer of oxygen.

Relatively few of the known dehydrogenases can use molecular oxygen as hydrogen acceptor however; the majority can communicate with oxygen only by way of cytochrome. Even of these only a minority can catalyse the transfer of hydrogen from substrate to cytochrome. Of those that can do this (Type 2) we may take succinate dehydrogenase as an example, and even in this case a flavoprotein is involved in the complete system. The other dehydrogenases (Type 3) require the co-operation of coenzymes as well as flavoproteins as *intermediate hydrogen carriers* through which the hydrogen is transferred, step by step, until at the end of the chain cytochrome is reached. We do not know why cytochrome can be reduced only indirectly and not directly; presumably the factor of enzyme specificity is in some way involved. To take a concrete example, the lactate dehydrogenase of muscle requires as intermediate hydrogen carrier a substance known as *coenzyme I*. This is identical with the so-called cozymase of yeast and is a dinucleotide of adenine and nicotinic amide (see p. 76) abbreviated as NAD. This coenzyme is required by most of the dehydrogenases of Type 3 and is very important in cellular respiration. A few members of this group, e.g. hexosemonophosphate dehydrogenase, require a different coenzyme, *coenzyme II* (NADP) in place of coenzyme I, and although it differs from coenzyme I only in that it contains three phosphoric acid radicles per molecule instead of two, the two coenzymes are not mutually replaceable. Hydrogen is passed on from the coenzyme to another carrier. This is itself an enzyme; it is in fact a conjugated protein of which the prosthetic group is a flavin nucleotide identical with that which we have already mentioned as the prosthetic group of amino-acid oxidase.

It will probably have been noticed that flavin and nicotinic amide, neither of which occurs in any great concentration, play parts of very great importance in the processes which underlie cellular respiration. Neither can be synthesised, at any rate by the tissues of the mammals, and both must therefore be supplied ready made in the diet of such animals. Both are vitamins, in fact, and both have been recognised as components of the vitamin B_2 complex. Adenine too ranks high in importance, and but for the fact that it is a constituent of the nucleoproteins of all cells and is therefore unlikely to be absent from any but the most unnatural of diets, it too might have found a place on the list of vitamins. Apparently it can be synthesised by animal tissues.

As a rule it appears that the greater part of the respiration of cells is carried on by dehydrogenases of the types which work, whether directly or indirectly, through cytochrome. In many cases, however, the rate of respiration of cells and tissues can be increased by adding a reversibly oxidisable and reducible dyestuff such, for instance, as methylene blue. This dye can act as a hydrogen acceptor, and in its reduced form is spontaneously re-oxidised in the presence of oxygen and is therefore able to function as an *auxiliary hydrogen carrier*. Thus methylene blue can take the place of cytochrome *plus* cytochrome oxidase in cells and tissues which have been poisoned with cyanide, but in normal tissues the dye can act side by side with the tissues' own cytochrome.

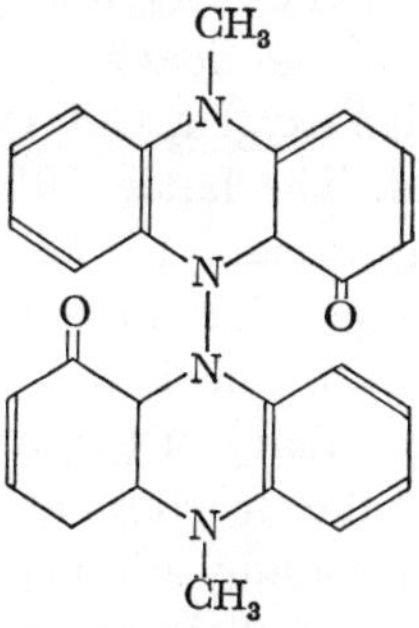

Pyocyanin

Methylene blue itself is not a naturally occurring compound but there exist several natural pigments which can act in the same way. The red blood cells of the rabbit and the eggs of the sea urchin are particularly sensitive to pigments of this kind. Very small concentrations of these substances, such as $M/5000$, are effective, and we must suppose that they are capable of acting as auxiliary respiratory catalysts for the cells in which they occur.

Four such pigments may be mentioned. Pyocyanin is a blue pigment produced by *Bacillus pyocyaneus*, and is a phenazine compound, while hallachrome, a red pigment found in the skin of the annelid worm *Halla*, is a quinonoid substance related to tryptophan but probably derived from tyrosine. Echinochrome, a red pigment found in the tissues of many echinoderms, and chromodorin, the purple pigment of certain molluscs, are two further examples.

4. *The transport of carbon dioxide*

The removal of carbon dioxide from the tissues involves a complicating factor that does not arise in the transport of oxygen for, unlike oxygen, CO_2 reacts with water to form carbonic acid. The latter, although a very weak acid, does ionise appreciably; it follows, therefore, that CO_2 transport is liable to lead to changes in the pH of the blood and tissue fluids. This is very important from the viewpoint of the constancy of the internal environment for it is known that the structural and catalytic proteins of living cells are profoundly affected by relatively trivial changes in the pH of their immediate environment. Large changes kill them altogether.

In its simplest form the problem of CO_2 transport can be represented by the following equilibria:

$$CO_2 + H_2O \underset{slow}{\rightleftharpoons} H_2CO_3 \underset{fast}{\rightleftharpoons} H^+ + HCO_3^-$$

These reactions tend to move towards the right in the tissues, where the partial pressure of CO_2 is high, and towards the left in the respiratory organs, where the partial pressure is low. The hydration of carbon dioxide by water is, unless catalysed in some way, a relatively sluggish reaction, a fact that is probably of considerable importance for slow-living animals. Such creatures produce little CO_2; little carbonic acid and few hydrogen ions are accordingly formed and CO_2 can be accommodated in solution in the blood and the tissue fluids with little danger that their pH will be appreciably disturbed. But in more active animals where large amounts of CO_2 have to be dealt with, higher partial pressures are likely to be built up, and more carbonic acid and more hydrogen ions are formed. In such cases the blood must have a high buffering capacity to deal with hydrogen ions, as well as a higher overall CO_2 capacity.

An important part in CO_2 transport is played by the blood proteins. These include respiratory proteins such as haemoglobin, haemocyanin and the rest, and other proteins that do not act as respiratory carriers of oxygen. As a rule the pH of the blood of animals is on the alkaline side of the isoelectric pH of the blood proteins which, in consequence, are negatively charged under physiological conditions. The blood proteins can accordingly react with hydrogen ions to form compounds which, for lack of a better title, can be collectively designated as 'proteinic acids':

$$H^+ + (Pr)^- \rightleftharpoons H(Pr)$$

Inasmuch as they yield hydrogen ions by their ionic dissociation these products are true acids, but they are so exceedingly weak that they scarcely dissociate at all; they are, in fact, almost perfect buffers. The more protein a blood contains, the more hydrogen ions it can accommodate without suffering any appreciable change of pH.

Although it is in their capacity as buffers that the blood proteins are mainly of respiratory importance, they contribute something more to the transport of carbon dioxide, which can combine with the free amino-groups of proteins to form 'carbamino-compounds':

$$-NH_2 + CO_2 \rightleftharpoons -NH.COOH$$

The following means of transporting CO_2 are thus available in protein-containing bloods:

(i) simple physical solution;
(ii) combination as carbonic acid;
(iii) combination as bicarbonate ions (this entails buffering of the corresponding hydrogen ions);
(iv) combination in the form of carbamino-compounds.

It is estimated that of the CO_2 transported in mammalian blood, 70–80 per cent travels in the form of bicarbonate ions, about 10 per cent in simple physical solution, and 10–20 per cent in the form of carbamino-compounds. Thus, as the blood circulates through metabolically active tissues the following reactions take place (full arrows), the whole series being reversed (broken arrows) in the respiratory organs:

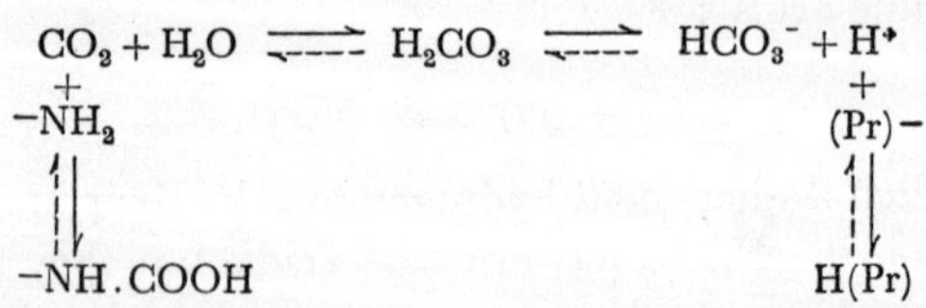

The part played by the blood proteins is clearly of the utmost importance in respiration, and this is probably true even of animals that possess no haemoglobin, haemocyanin or other respiratory carrier of oxygen. With the greater activity made possible by the possession of such a carrier, carbon dioxide production increases as the oxygen consumption increases, but the CO_2 capacity and buffering power of the blood are proportionately increased at the same time by precisely the material that makes the greater activity possible, the respiratory protein itself. Thus, in the animal kingdom as a whole, there is to be seen a general parallel between physiological activity and the general level of organisation on the one hand and, on the other, the oxygen capacity, CO_2 capacity, buffering power and protein content of the blood.

In many animals, especially in the more active types and species, there is present in the blood a highly specific enzyme known as carbonic anhydrase, which catalyses the following reaction:

$$CO_2 + H_2O \rightleftharpoons H_2CO_3$$

It is probable that carbonic anhydrase is involved in a number of other important processes, but its main function is concerned with respiration. In its absence the slow reaction between CO_2 and water to form carbonic acid, and the slow decomposition of the latter into its parent products, constitutes a bottleneck that might delay the uptake of CO_2 from highly active tissues and its subsequent discharge at the respiratory organs. Carbonic anhydrase has the effect of widening the bottleneck so that CO_2 entering the blood corpuscles is rapidly converted into carbonic acid, the ionic dissociation of which is a virtually instantaneous process. The subsequent reaction between hydrogen ions and protein ions is likewise almost instantaneous so that, in the presence of the enzyme, CO_2 can be accommodated with great rapidity in the blood and re-formed equally rapidly for discharge at the respiratory organs.

Unfortunately we do not know very much about the distribution of this important enzyme. Most is known about the part it plays in the bloods of vertebrates, especially of the mammals. Here, it will be remembered, the blood plasma, itself rich in proteins (albumins and globulins), contain enormous numbers of red corpuscles, each of which is rich in haemoglobin. Although it is certain that some CO_2 is carried in the form of carb-amino-compounds with the plasma proteins, the bulk is transported within the corpuscles. *Carbonic anhydrase is exclusively confined to the corpuscles and none is present in the plasma.* Little carbonic acid can therefore be formed in the plasma, but CO_2 diffusing into the corpuscles is promptly converted into carbonic acid. The latter ionises in the usual way, the resulting hydrogen ions are buffered on the haemoglobin, and bicarbonate ions pile up inside the red cells.

Before we follow up the further implications of these changes it is well to notice that, because carbonic anhydrase is confined to the corpuscular contents, the formation of hydrogen ions is almost completely confined to the contents of the red cells; few if any are formed in the plasma, which constitutes the true internal medium of the mammal. The red cells, therefore, may be regarded as constituting a specialised respiratory tissue—a diffuse, circulating tissue as contrasted with the compact masses of cells that make up the 'fixed' tissues of the body—but a tissue none the less, for it is within these cells that oxygen and carbon dioxide alike are transported, and it is in these cells, removed as they are from the true *milieu intérieur*, that the dangerous but necessary process of hydrogen ion-formation takes place.

The changes that take place when a red cell loads or unloads its burden of carbon dioxide is reflected by changes in the plasma, but these are not such as to have any noxious effects upon the tissues. When a red cell loads up with CO_2, the bulk of the latter is rapidly converted into bicarbonate and there is a steep rise in the concentration of bicarbonate ions within the cell. The corpuscular membrane is permeable to these ions, which accordingly begin to diffuse away into the plasma. If this process were to continue, however, the corpuscles would become positively charged and it might be supposed that their electrical neutrality would be maintained by allowing positively charged ions to leave along with the negatively charged bicarbonate ions. This, however, is not the case, for the corpuscular membrane is impermeable to positively charged ions and particles. Electrical neutrality is maintained nevertheless and this is accomplished by the inward migration of chloride ions, which are abundant in the plasma. When, therefore, a red cell loads up with CO_2, bicarbonate ions leave the corpuscle and are replaced by chloride ions drawn from the plasma; this exchange constitutes the 'Hamburger phenomenon', otherwise known as the 'chloride shift', which is reversed later when the cell yields up its CO_2 in the respiratory epithelia. The whole sequence of events leading up to the 'chloride shift' is summarised in Fig. 12.

One interesting consequence of the Hamburger phenomenon is that, as we pass from less to more active animals, there is a marked decrease in the chloride content of the internal medium (see Table 3, p. 15), the decrease being compensated by a corresponding increase in bicarbonate content. We have here another example of the fitness of the environment to which some attention

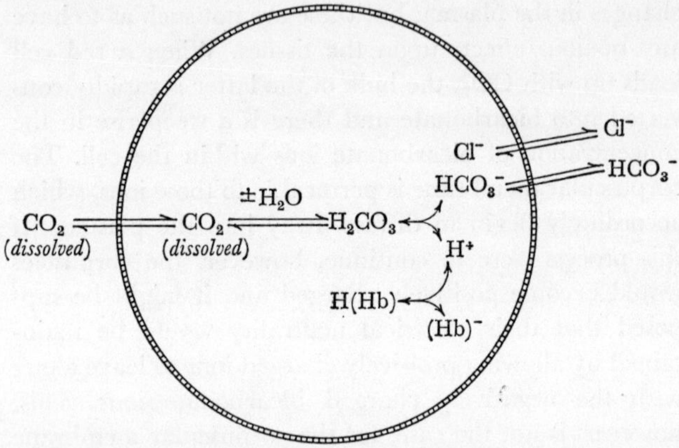

Fig. 12. To illustrate the mechanisms involved in CO_2-transport and buffering in the mammalian red blood cell. The corpuscular membrane is impermeable to positively charged ions. Carbonic anhydrase is confined to the red cells; none is present in the plasma.

has already been given (p. 22). As it happens, the physiological properties of Cl^- and HCO_3^- are virtually identical. Were this not so, it is doubtful in the extreme whether the mechanisms that lead to the 'chloride shift' could ever have been put to biological service, in which case the transport of carbon dioxide on the grand scale could not have been accomplished in this particular manner and, indeed, might never have been achieved at all.

In conclusion, some mention must be made of what is called the Bohr effect, which is concerned with the influence of pH upon the pigment's affinity for oxygen. The hydrogen ions formed when a red cell loads up with carbon dioxide are accommodated on the haemoglobin so that the charge on the latter becomes slightly less negative. This has the effect of somewhat reducing the affinity of the pigment for oxygen. Consequently, when a red cell passes through a metabolically active tissue and becomes loaded up with carbon dioxide, its oxyhaemoglobin dissociates rather more freely than it would otherwise do. The production of carbon dioxide therefore facilitates the provision of oxygen to the tissues. In the respiratory organs the effect is reversed: as carbon dioxide is discharged, hydrogen ions leave the haemoglobin, which becomes more negatively charged and accordingly takes up oxygen more readily.

This mechanism is obviously of considerable respiratory value and has been thought by some to be of adaptational significance. This, however, is doubtful, for in many animals the Bohr effect is reversed, i.e. the addition of hydrogen ions to the blood makes the dissociation of the oxygenated pigment more rather than less difficult. Thus, while the 'direct' Bohr effect observed in mammalian blood is probably of considerable respiratory value, it is difficult to believe that it has an adaptational significance. We must rather believe that this particular property of the respiratory pigments happens fortuitously to operate to the advantage of some animals, but that it operates detrimentally in animals where the effect is reversed.

CHAPTER VII

1. *Animal pigmentation*

THE colours of animals have long had an irresistible fascination for biologists. Zoologists speak of the 'protective' coloration whereby an animal so closely resembles its natural surroundings as to be detectable only with difficulty and thus enjoys relative immunity from predatory attacks. Many animals, of course, are sufficiently well armed with one or another kind of biological weapon to have no need for camouflage, and some, indeed, are so boldly coloured as actually to advertise their presence. It is thought that coloration of this kind can act as a warning to would-be predators that any attack will have unpleasant consequences, and hence it is usual to speak of 'warning' coloration in this connection. Other animals, less innocuous in themselves, appear to secure a considerable degree of safety from their enemies by 'mimicry' of the so-called warning coloration of other more offensive creatures.

The usefulness of animal coloration depends upon the visual range of the observer. Carnivorous animals are mostly—perhaps universally—colour-blind and unlikely to perceive the warning colours of a wasp or a hornet. Again, colour as such can be of little interest to deep sea animals or to animals that mate in the dark, but in such cases colour is often replaced by fluorescence. Furthermore, proteins appear colourless to mammals, unless they happen to possess a chromophoric prosthetic group.

Colourless proteins do however have strong absorption bands in the ultraviolet and must therefore look black to insects that can 'see' in this region.

Closely interwoven with colour is pattern. Often it is difficult to decide which, if either, of the two is of the greater survival value to its possessor, and the whole field of animal coloration, from this point of view alone, bristles with unsolved problems. But coloration is interesting too from other points of view. Beauty is by no means always only skin deep, and the significance of internal colorations can hardly be discussed in terms of the external appearance of the animal. Many of the animals whose blood is red show no trace of externally visible redness, yet the red blood pigment haemoglobin is of absolutely vital importance to them.

It is a curious fact that many substances of the first order of metabolic importance are coloured. Colourless proteins frequently carry coloured prosthetic groups; thus the proteins of the haemoglobins, haemocyanins, chlorocruorins, cytochrome and so on are all colourless and appear red, blue, green or red again when, and only when, they are combined with the appropriate prosthetic group.

Interestingly enough, the chemical properties that go to the production of colour are also associated with oxidation-reduction processes such as we see in the behaviour of cytochrome, or with the formation of complexes with oxygen such as oxyhaemoglobin, oxyhaemocyanin and the rest.

Respiratory carriers of oxygen, such as haemoglobin and haemocyanin, respiratory catalysts, such as cytochrome, and even certain enzymes, such as catalase, peroxidase, amino-acid oxidase and xanthine oxidase among others, are all coloured, but many of them occur in such minute quantities that their coloration, as such, can scarcely be of any value, positive or negative, to the owner. Moreover, they occur as a rule in the deep tissues rather than in the integument.

It would thus seem that colour is in many cases a chemically inevitable concomitant of metabolic activity. Perhaps it is not unreasonable, therefore, to expect that many coloured compounds will, when we know more about them, turn out to be of metabolic significance, whether or not their colour is of any importance in its own right. Animal pigments, then, are interesting from the metabolic point of view, and it is from this view-point that the following sections have been written. If this chapter seems, as a whole, to be nothing more than a catalogue of unrelated facts, it is because we do not yet know enough about pigments and their metabolism to build up any coherent sort of picture from the material at present available.

2. *Pyrrol pigments*

We have already discussed some of the pigments which contain the pyrrol ring, viz. haemoglobin, chlorocruorin, helicorubin and the cytochromes. The haem nucleus which is present in all these consists essentially of four pyrrol rings joined together as in the skeleton formula opposite. An iron atom is present and side chains are attached at positions 1–8. Chlorophyll too contains a nucleus of this kind but the iron atom is replaced by one of magnesium in this case and the side chains are different.

If the atom of metal is removed, we are left with tetrapyrrol compounds belonging to the group known as porphyrins. These are very widely distributed and differ from each other mainly in the nature of the side chains. *Protoporphyrin* is the name given to the porphyrin derived from haemoglobin without modification of the side chains. The better-known *haematoporphyrin* also can be

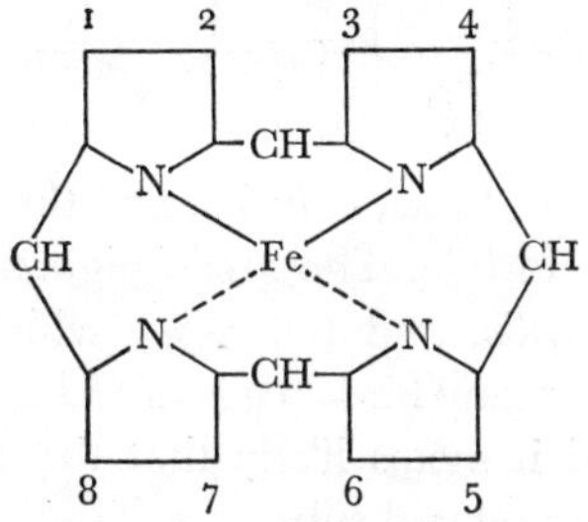

obtained from haemoglobin, but differs from protoporphyrin in that the side chains are modified. *Aetioporphyrin* is a product derived from chlorophyll, and is interesting because it can also be obtained from haematoporphyrin by treatment with soda-lime, showing that haemoglobin and chlorophyll have a closely similar basic structure in spite of their widely different functions. *Ooporphyrin*, a pigment which occurs in the shells of many birds' eggs, is identical with haematoporphyrin. *Coproporphyrin* and *uroporphyrin*, both of which are derived from haemoglobin, appear in the urine of human patients in cases of a rather rare genetic disorder known as porphyrinuria; they are also deposited in the bones in such cases, colouring them a dark brown. *Turacin*, the magnificent red pigment of the feathers of the turaco, a South African bird, is a copper derivative of uroporphyrin, while *conchoporphyrin*, the pigment of the shell of the pearl mussel, is a carboxylated coproporphyrin.

Another group of pyrrol pigments is derived from the porphyrins by opening up the ring and removing the iron. These, the bile pigments, have the following skeleton formula:

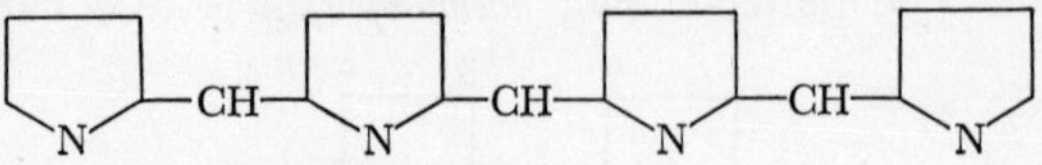

The best known of these, *bilirubin*, and the closely related *biliverdin*, occur in bile. The green pigment of the dog's placenta, *uteroverdin*, and the *oocyan* which imparts the green colour to many birds' eggs are identical with biliverdin. Indeed it seems likely that these two last pigments and many related substances found in the cuticles of many invertebrates are just convenient ways of disposing of haem materials that are of no further use to their owners. All these pigments contain four pyrrol rings.

The vanadium chromogen of the Tunicata appears to be a pyrrol compound, but we do not know how many rings it contains. A tripyrrol compound, *prodigiosin*, is produced by *Bacillus prodigiosus*, an organism which grows with great readiness on damp bread. The pigment is bright red in colour and is thought to have been responsible for the well-known medieval miracles of the Bleeding Host, which provided an apparent proof of the doctrine of transubstantiation in Latin theology. Apart from its habit of performing miracles the function of this pigment is entirely unknown: its formula shows that it is not directly related either to the porphyrins or to the bile pigments.

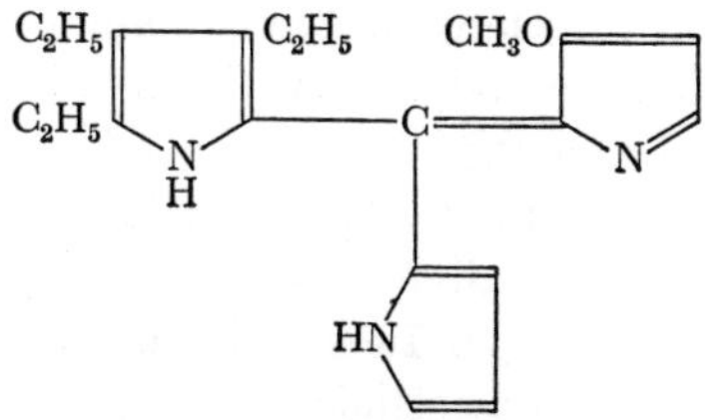

Prodigiosin

3. *Indol pigments*

The widely distributed melanins are black pigments which contain the indol ring. Although they resemble tryptophan in the possession of this ring, they are actually formed from phenolic substances including tyrosine through the action of the oxidising tyrosinase. The following are some of the stages involved:

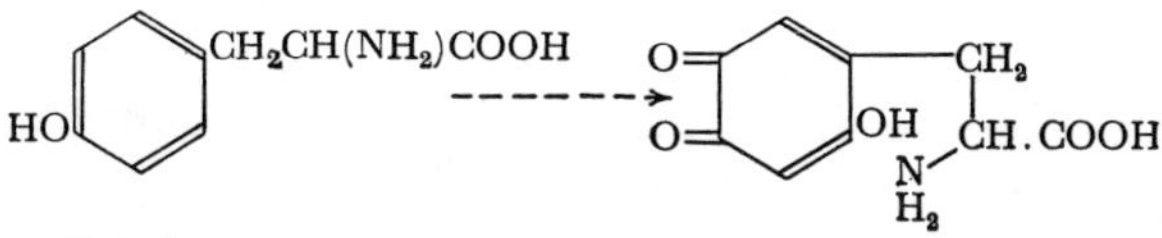

Tyrosine

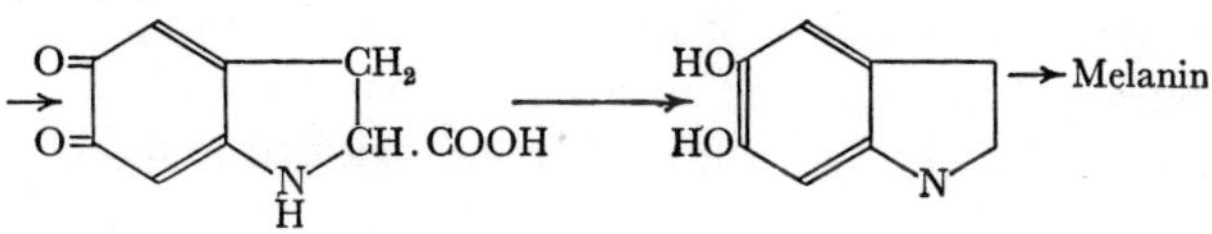

Red intermediate compound — Polymerisation

In black-and-white animals such as piebald rabbits and guinea-pigs, tyrosinase is present in the black parts of the skin and absent from those which are white, albinism being associated with the complete absence of tyrosinase from the body. In variegated insects such as the spurge hawk moth, *Deilephila euphorbiae*, on the other hand, it seems that the localisation of the black pigment is associated with localisation of the substrate and not with that of the enzyme. In some insects (e.g. cockroach) the substrate is protocatechuic acid rather than tyrosine. Reference has already been made (p. 62) to the pterines, another group of insect pigments.

The black or dark brown 'ink' of the octopus and cuttlefish is also a melanin. It is formed in a special gland, the ink-sac, and if the animal is alarmed it puffs out a great cloud of the finely divided pigment which, coming between the animal and its pursuer, serves the purpose of a smoke screen, while the animal, taking advantage of the situation, darts away in a different direction. In passing it is interesting to notice another aspect of tyrosine metabolism in the Cephalopoda. The bite of the octopus has been said to owe its poisonous nature to tyramine, a compound which probably arises by simple decarboxylation of tyrosine.

Indigo, the natural colouring matter of woad, is another indol compound. The woad plant, *Isatis tinctoria*, contains a glycoside of indoxyl, and when the juice is extracted and exposed to air the sugar molecule is removed, indoxyl being liberated and oxidised to indigo:

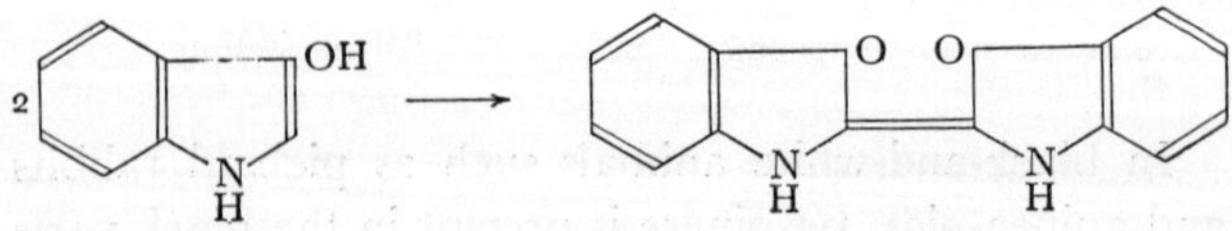

Indoxyl

Indigo

Another natural dye, Tyrian purple, is a brominated indigo. It was known to the ancients, who used it to dye cloth for senatorial and imperial robes and made it by crushing the viscera of the marine gastropod *Murex*, and exposing them to air and sunlight. Tyrian purple is formed in much the same way as indigo itself, except that its precursor is a sulphur compound instead of a glycoside. A similar and probably identical pigment can readily be prepared from the common dog whelk, *Purpura*.

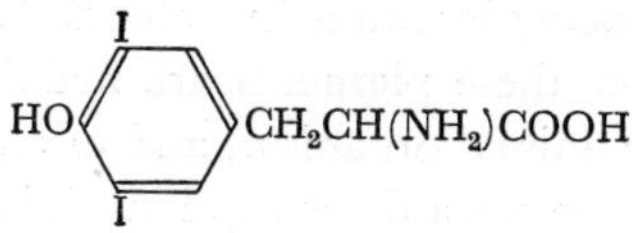

Iodogorgoic acid

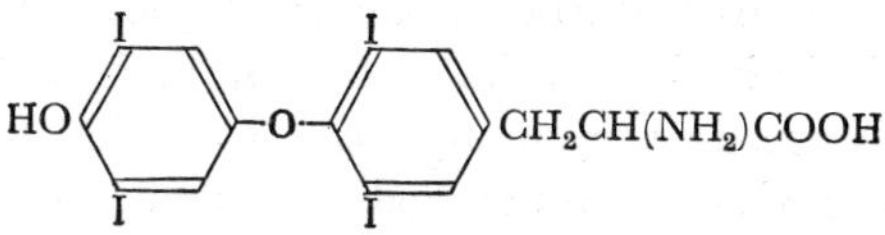

Thyroxin

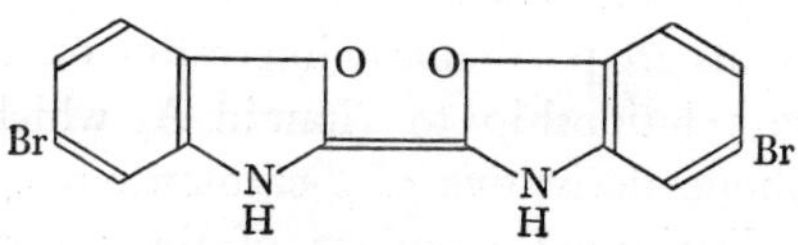

Tyrian purple (*6-6′*-dibromindigo)

It is convenient to refer at this point to another brominated compound, bromogorgoic acid (dibromo-tyrosine) which together with iodogorgoic acid has been obtained by hydrolysing the skeletal substance of a coral, *Gorgonia*. Both are halogenated tyrosines, and it is interesting to recall the occurrence of iodogorgoic acid in thyroid tissue, and its evident relationship to thyroxin itself.

4. *Carotenoid pigments*

Carotenoid pigments are among the most generally distributed of natural pigments. They occur especially widely among plants, from which they are transferred to the animals. Being insoluble in water and soluble in fats and fat solvents, these pigments are mainly associated with fats in the animal organism, and of all the familiar animal fats none, with the exception of pig fat, is wholly free from yellow carotenoid materials. The carotenoids which occur most commonly in animal tissues are *carotenes*, which are hydrocarbons, and *xanthophylls*, which are alcohols derived from the carotenes. The carotenoids of animals are, in very many cases, unmodified pigments directly derived from the food, but often enough, especially among invertebrates, they are substances which, though derived in the first place from the food, have undergone more or less extensive chemical modification, apparently in the tissues of the animals themselves.

These pigments are particularly interesting on account of their close relationship to vitamin A, which is in itself an alcoholic derivative of β-carotene, one of the most widely distributed members of the group. The skeleton formulae shown opposite show the relationships between α- and β-carotenes and vitamin A; the carbon atoms marked with an asterisk correspond to each other and this helps to show how one molecule of β-carotene can give rise to two of vitamin A, whereas one molecule of α-carotene gives rise to only one of the vitamin. It has been discovered that a second vitamin A (A_2) occurs in fresh-water fishes: it closely resembles A except that the ring contains two double bonds instead of one. In euryhaline, migratory fishes, A and A_2 occur together, side by side.

As is well known, vitamin A deficiency has as one of its consequences the condition known as night blindness. The relation of this disorder to the avitaminosis has now become clear, for visual purple is known to be a conjugated protein, the prosthetic group of which is the aldehyde corresponding to vitamin A. A vast amount of work on visual pigments and the so-called visual cycle has been done over many years, pioneered by Wald, and the occurrence of essentially similar substances and processes has been demonstrated in a variety of invertebrates.

Of the xanthophylls, lutein and zeaxanthin are particularly common in plants and animals alike. These are dihydroxy derivatives of α- and β-carotene respectively. Both are found in the yolks of birds' eggs for example.

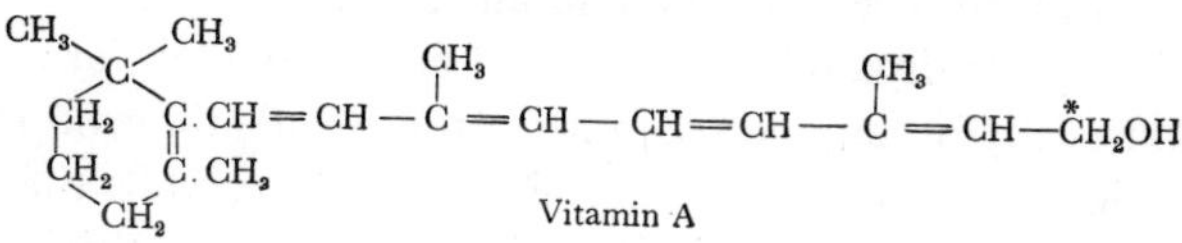

Vitamin A

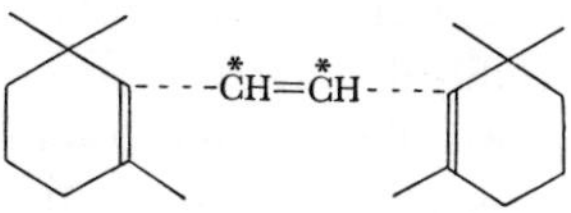

β-carotene

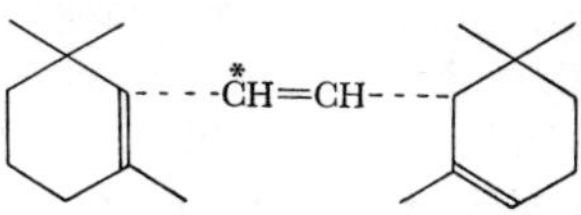

α-carotene

By feeding hens on diets containing no xanthophylls, eggs with colourless yolks can be obtained. The colour is not restored if carotenes are added to the diet, and from this observation it may be deduced that the avian organism cannot introduce (OH) groups into the molecules of the carotenes. Nevertheless, given xanthophylls, some birds are capable of modifying them to produce special xanthophylls of their own. Thus the yellow pigment of canary feathers (which is found, by the way, in the feathers of a number of other birds as well) is a special xanthophyll formed from the food xanthophylls, but one which cannot be formed from carotenes. A number of other special 'animal xanthophylls' are known, and among them we might mention the brilliant red pectenoxanthin of the ovaries of the scallop, *Pecten*.

Another of these animal xanthophylls, astaxanthin, is very widely distributed among invertebrates, especially Crustacea. This is a diketo-dihydroxy-β-carotene which easily undergoes oxidation under certain conditions to give a tetraketo-β-carotene, astacin, in which form it has long been known under a variety of different names. The red hypodermis of the lobster *Homarus* and a number of related forms owes its colour to an ester of astaxanthin, probably the dipalmitate. The blue shell of the lobster is coloured by a conjugated protein, the prosthetic group of which appears to be astaxanthin dipalmitate, while the green eggs contain a similar pigment, the prosthetic group in this case consisting of unesterified astaxanthin. Both these chromoproteins are readily denatured by heat, alcohol and the like, and the prosthetic groups, both of which are red in colour, are set free. This accounts for the familiar red coloration of the lobster after cooking.

Astacin, presumably derived from astaxanthin, has been obtained from a wide range of Crustacea, large and small, as well as from a number of other invertebrate forms, while the pink colour of salmon muscle, from which astacin has been isolated, testifies to the partiality of this fish to the many small crustaceans which are included in its diet. So much work has been done in this field over the years that this list of animal carotenoids might be much extended, but the foregoing brief account must suffice for present purposes.

5. *Animal luminescence*

The luminescence of the glowworm is too well known to call for any description, but it is worth while to notice that the phenomenon of luminescence has been observed in upwards of forty orders of animals, as well as in numerous bacteria and fungi. The phosphorescence of the sea, the ghostly appearance of rotten tree stumps and the 'glow' of stale meat and fish in the dark are all due to one or another of these organisms. The light comes from fine granules which may glow within the cells themselves or may be extruded and continue to glow even after the animal has taken its departure.

The most thoroughly studied case is that of a small crustacean, *Cypridina*. Here the cells contain granules of two kinds, some large and yellow in colour and the rest small and colourless. The large yellow granules consist of a dialysable substance, which dissolves in alcohol, is not attacked by trypsin, and is therefore unlikely to be a protein, though we still do not know its precise chemical nature. It is known as luciferin and is thought to be related to naphthohydroquinone. The small colourless granules consist of a non-dialysable substance which is rapidly destroyed by trypsin, insoluble in alcohol, and precipitated from its aqueous solution by protein precipitants. This substance, luciferase, is of protein nature, and is a cyanide-sensitive, oxidising enzyme, its characteristic behaviour being destroyed by heat. When luciferase is mixed with luciferin, light is produced, and it has been shown that the luciferase catalyses the oxidation of luciferin to a dehydroluciferin. In common with most enzymes luciferase is very specific, and enzyme prepared from a given species will only act on the luciferin of the same or a closely related species.

The oxidation can be brought about in other ways, as, for example, by the use of potassium ferricyanide, but no light is then produced. It is therefore believed that it is the enzyme and not its substrate which gives out the light, and the following reactions appear to be involved:

(1) Luciferin $+ \frac{1}{2}O_2 \rightarrow$ oxyluciferin$'$ $+ H_2O$.
(2) Oxyluciferin$'$ + luciferase $\rightarrow$ oxyluciferin + luciferase$'$.
(3) Luciferase$'$ $\rightarrow$ luciferase $+ h\nu$.

The first reaction is catalysed by luciferase and the product, oxyluciferin, carries a quantum of energy represented in the equations by a prime. This quantum of energy is transferred to the enzyme itself in the second reaction and is liberated, in the third reaction, as a quantum of light energy, $h\nu$.

CHAPTER VIII

1. *Nutrition*

Animals are not always struggling for existence. They spend the greater part of their time doing nothing in particular. But when they do begin, they spend the greater part of their lives eating. Feeding is such a universal and commonplace business that we are inclined to forget its importance. The primary driving force of all animals is the necessity of finding the right kind of food and enough of it.

CHARLES ELTON

NUTRITIONAL studies of animals have mainly been concerned with mammals, but there is little reason to think that animals of other kinds differ very markedly from them, for a review of the rather scanty evidence at present available leads to the conclusion that the metazoan invertebrates which have been studied have nutritional requirements which are at least as complex as those of the vertebrates. Chemical analysis of foodstuffs of many different kinds has shown that, apart from water, they consist mainly of proteins, fats and carbohydrates. An enormous number of feeding experiments have gone to show that proteins are an indispensable article of food for mammals. Of the twenty-odd amino-acids that go to make up a typical protein, about ten or a dozen are compounds that cannot be synthesised by the animal itself and, for these 'essential' or indispensable amino-acids, the animal has to rely upon the proteins of its food. There is also evidence that certain fatty acids are similarly indispensable but no reason at present to think that the carbohydrates of the food include any indispensable dietary constituents.

In addition to the essential amino- and fatty acids,

however, animals have to rely upon their food for provision of a considerable number of vitamins. These, like the other essential substances, the animal must have, but cannot synthesise for itself. There can be no doubt whatever of the importance of these 'essential' dietary factors, if only because animals sicken and die if deprived of any single one of those they require. In some cases we know at least some of the precise functions that these compounds discharge. Amino-acids, for example, are required as building stones for the construction of new tissue proteins during periods of growth, and those that the animal cannot produce for itself it must obtain ready-made in its food. Amino-acids are similarly needed for the day-by-day elaboration of special secretions such as enzymes and hormones and for a number of other special purposes. The vitamins likewise are indispensable constituents of important compounds. Coenzymes I and II (NAD and NADP) and the flavoproteins, which play fundamental parts in oxidative metabolism, contain nicotinic amide and riboflavin respectively, and these are two members of the vitamin B_2 complex. Co-carboxylase, another substance that is intimately involved in many metabolic processes, is the diphosphate of vitamin B_1 (thiamine), while normal vision involves the participation of visual purple, a derivative of vitamin A, and so on. Certainly it cannot yet be claimed that we know all the functions of all the vitamins and essential amino-acids, but of their importance in tissue structure and metabolism we may feel quite confident.

All these essential substances can be produced by green plants, from which herbivorous animals collect them directly; carnivorous beasts also obtain their supplies from the green plants, though at second or third hand. Animals of every kind are therefore dependent upon the green plants, not only for their supplies of energy-yielding foods such as fats and carbohydrates, but for their indispensable accessory food factors as well.

Green plants, unlike animals, are able to synthesise everything their life requires from very simple raw materials. The energy required for these synthetic achievements is obtained from the inorganic world by trapping solar radiation with the aid of chlorophyll. Provided that supplies of water, carbon dioxide, salts, and a simple source of nitrogen such as ammonia or nitrate are available, the green plant can accomplish the synthesis of all the amino- and fatty acids, proteins, fats, carbohydrates and vitamins, and a wealth of other substances, simple and complex. Photosynthesis is, indeed, the most important of all biochemical operations and it is a matter of great satisfaction that, after many years of intensive study, biochemists have been able to lay bare many details of its underlying mechanisms. These mechanisms, unfortunately, are too complex to be discussed in these pages.

It is clear that animals are much inferior to plants in synthetic ability and, since there must presumably have been a time when all the living inhabitants of the world were self-supporting, it becomes interesting to enquire how the animals have come to lose so large a part of the synthetic ability that their ancestors must at one time have possessed. Even the simplest and most primitive of modern animals appear to have nutritional requirements—and therefore synthetic disabilities—that do not differ strikingly from those of the modern mammals, and it seems likely therefore that their loss of synthetic competence must have taken place at an early date in their evolutionary history.

The green plants of the present day can synthesise complex organic materials when light is available. They can also, during the hours of darkness, obtain the energy necessary for their continued existence by breaking down complex materials previously synthesised in the light. If we suppose that the earliest plant-like organisms similarly possessed the ability both to synthesise and to break down organic complexes, we can imagine that some of them may have lost their former photosynthetic abilities and thenceforth became dependent upon other organisms for supplies of energy-containing food. Something of this kind has led to the development of the modern fungi. The fungi do not photosynthesise, but rely on the availability of organic matter from which, by chemical decomposition, they can obtain the energy necessary for the accomplishment of their own synthetic operations.

But there remains a large gap between organisms which, like most fungi, can synthesise everything they need so long as a suitable organic energy-source is available, and the animals, which require many accessory food factors over and above mere energy-yielding organic foods. This gap is filled in by numerous micro-organisms, especially by bacteria, and it is to some of these that we must next turn our attention. Although at first glance these unicellular organisms may appear to be utterly dissimilar and quite unrelated to animals, it turns out that the nutritional requirements of the two groups are remarkably similar. The essential amino-acids and many of the vitamins required by animals are also required by one or another of the many known types and species of micro-organisms, and recent work on their nutrition has gone a long way towards enabling us to account for the nutritional requirements of animals.

Among the bacteria and protozoa there are some that are self-supporting like the green plants, i.e. they are *autotrophic*. Some are *photosynthetic* and are usually coloured purple or green by the presence of bacterial counterparts of chlorophyll. Others are *chemosynthetic*; these, like the photosynthetic organisms, are autotrophic, but do not utilise solar energy. Instead they are able to harness to their own synthetic purposes the energy liberated in simple inorganic reactions taking place in the external world. Some obtain energy from the oxidation of elemental sulphur to sulphuric acid, for example, others from that of ammonia to nitrite or nitrate, while yet others exploit the oxidation of ferrous to ferric compounds. The nutritional requirements of all these autotrophes, plants and micro-organisms alike, are very simple indeed. Water and salts are needed as, indeed, they are by all living organisms; carbon is obtained from carbon dioxide, nitrogen from some simple inorganic source, while the necessary energy is collected from the inorganic world outside. Organic sources of energy are not required and, indeed, actually inhibit the growth of many autotrophic bacteria.

The remaining organisms differ from the autotrophes in that they have no power to acquire energy directly from the inorganic world, but must be provided with organic substances by the fermentation or oxidation of which the energy necessary for their synthetic operations can be obtained. Such organisms are said to be *hetero-*

trophic. Heterotrophic organisms in general can be further subdivided according as they are or are not '*exacting*'.

Organisms which are not 'exacting' can live on simple media containing water, salts and a simple nitrogenous substance, together with some suitable organic compound such, for instance, as glucose or lactate. Non-exacting organisms of this kind are usually free-living; they can settle down wherever their simple food-requirements can be satisfied and can then elaborate all their tissue constituents and catalytic apparatus without further aid from their environment. But many other micro-organisms cannot grow under such rough and ready conditions and can grow and multiply only if one or more special substances are available to them, so that they are necessarily restricted to habitats of particular kinds. Such organisms are said to be exacting towards the particular substance or substances that form their indispensable accessory food factors. There are many degrees of exactingness. Some organisms are content with only one or two special factors; others require many. At present the championship for exactingness probably goes to *Streptococcus haemolyticus*, some strains of which require at least 19 amino-acids and a considerable group of vitamin-like 'growth factors' into the bargain! Since these accessory food factors are indispensable it must be supposed that the organism in question cannot synthesise them, and it may justifiably be said that the exactingness of a given organism is a measure of its synthetic disability.

We may now return to enquire into the causes of the loss of synthetic ability that is so marked a feature of animals and many other heterotrophes. Free-living heterotrophes can settle down and make a living wherever the necessary water, salts, nitrogenous and organic compounds are to be had. Many free-living bacteria, therefore, can live in milk, which provides them with everything they require. Now it happens that milk is a fairly rich source of riboflavin, a compound which enters into the constitution of certain flavoproteins intimately involved in cellular metabolism. Free-living forms can produce this substance for themselves, but it has been found that many of the organisms that habitually live in milk are incapable of synthesising riboflavin; they cannot grow except in media that already contain this important substance. Probably, therefore, it may be argued that these milk-sourers were once free-living organisms that settled down and lived for so many generations in milk that they became accustomed to finding a ready supply of riboflavin in their habitual environment, so that their ability to synthesise it was lost for lack of employment.

It appears that the loss of synthetic ability does not take place suddenly and all at once, but slowly and in finely graduated stages. Thus *Proteus vulgaris*, given nicotinic acid or its amide, is capable of amidating the free acid if necessary and elaborating the adenine-nicotinic amide dinucleotides (NAD and NADP), which it requires as coenzymes for its metabolic machinery. In other organisms, in *Haemophilus influenzae* for example, synthetic ability has declined still further, and neither the acid nor the amide allows growth and multiplication; only nicotinic amide nucleotide or complete molecules of these coenzymes can satisfy its demands. Similarly, some strains of the diphtheria organism, *Corynebacterium diphtheriae*, are known to require the special and peculiar amino-acid, β-alanine, apparently as a raw material for the manufacture of pantothenic acid; this, which is another member of the B_2 group of vitamins, contains the β-alanyl radical. Other related strains of the same organism are unable to utilise β-alanine and must be given pantothenic acid itself.

With the facts we have considered in mind, it now begins to be possible to give a plausible account of the evolution of the synthetic disability that is so marked a feature of animal metabolism. Supposing that the world was at one time peopled only by self-supporting, autotrophic organisms, either chemo- or photosynthetic, we may imagine that some individuals lost their power to obtain energy from the external, inorganic world, but retained, as the fungi have done, the ability to break down organic complexes, thus giving rise to the first heterotrophes. As yet these new heterotrophes were non-exacting. Perhaps some early fungus-like organism developed the ability to capture and digest particles of dead or decaying stuff derived from their contemporaries and thus gave rise to *Amoeba*-like creatures, and thence, as new mechanisms for the capture, digestion and assimilation of food were evolved, to the rest of the present-day animal kingdom. Becoming accustomed to the consumption of organic matter, these organisms, from the earliest days in their evolution, must have found ready-made in their food a number of chemical compounds which formerly they had been obliged to manufacture for themselves but for which they could now afford to rely on their food, becoming more and more exacting with time. One of the most recent stages in the increasing exactingness of animal types is, perhaps, the loss of the ability to synthesise ascorbic acid (vitamin C), for among the mammals this compound is only required as an accessory food factor by the higher apes and by the guinea-pig, all other animals so far studied being able to synthesise this 'vitamin' for themselves.

However fanciful the notions of the last paragraph may appear or prove to be, the facts upon which they are based have indubitable value from the ecological standpoint. It is well known to biologists that plant and animal communities consist essentially of interlinked food-chains, each of which starts with autotrophic organisms (usually green plants) and proceeds through herbivores to carnivores, thence through smaller and smaller numbers of larger and larger carnivores until we arrive at a carnivore so large and powerful that it has virtually no enemies excepting man, old age and, perhaps, rapacious parasites. This social organisation, as we can now appreciate, is very largely based upon nutritional requirements, and we can observe how, in food-chains of every kind, the heterotrophic animals are dependent in the end upon autotrophic organisms. But biologists are also acquainted with other, less straight-forward associations, two of which stand out for particular biochemical enquiry, namely, symbiosis and parasitism.

Symbiosis

Many cases of symbiosis have been described from time to time, but only in recent years has this fascinating problem received much biochemical attention. In one common form of symbiosis, the occurrence of unicellular green algae within the cells of animals is seen. One well-known example of this type is found in the small platyhelminth, *Convoluta roscoffiensis*, which occurs in enormous numbers on the sands at certain places on the coast of Brittany. *Convoluta* hatches from the egg as a perfectly ordinary looking flatworm, possessing an alimentary canal of the kind usual among flatworms, and devoid of the green Chlamydomonads. It soon loses its digestive apparatus, however, and becomes infected with the symbionts, dying if this infection is prevented. From the time of infection onwards the two organisms live happily together in partnership until, for some reason at present unknown, the worm digests the algae and proceeds to die. This case seems to have been one of the first to be studied, and Keeble thought that the algae utilise waste nitrogen from the flatworm's metabolism (presumably ammonia), giving products of their own photosynthetic activity in return. Further study of the problem with the aid of modern biochemical methods would certainly be well rewarded.

Symbiosis is particularly common in herbivorous and wood-eating animals. Enormous numbers of animals, from insects to elephants, derive a great deal of their nourishment from cellulose, yet few of them are capable of digesting it. Cellulases have been described in a number of herbivorous molluscs, e.g. *Helix*, *Strombus* and *Aplysia*, in the wood-eating shipworm, *Teredo* (also a mollusc) and more recently in silver-fish, but the occurrence of such enzymes in animals appears to be rather rare. Cows, sheep and other familiar herbivores possess no cellulase and delegate the task of digesting cellulose and similar cow-indigestible polysaccharides to vast hordes of symbiotic micro-organisms which they accommodate in a specialised and commodious dilatation somewhere in the alimentary tract. Thus in the sheep, for example, we find that one of the four so-called 'stomachs', the rumen, and in the horse the caecum, are specialised for the reception and entertainment of symbionts. The rumen contents of the sheep carry out a rapid breakdown of cellulose and other carbohydrates, yielding large volumes of gaseous products, together with acetic, propionic, butyric and traces of other fatty acids. Thus the symbionts receive a liberal diet and ideal conditions for their own multiplication, in return for which they present the host with materials from which carbohydrate and fat can be synthesised. Essentially similar processes have been shown to occur in the ox, red deer, horse, pig, rabbit and rat.

But it is certain that the services rendered by the symbionts do not stop here. Although symbiotic and constrained to reside in some specific host, these organisms, like free-living yeasts and bacteria, appear to have the power to build up anything and everything they require for their own maintenance and growth. From what has gone before, the reader will realise that these substances include essential amino-acids and many, if not all, of the vitamins required by the host itself. Now even micro-organisms are not immortal and, when eventually they die, their tissues undergo autolysis and digestion with liberation of the constituent amino-acids, vitamins and so on, all of which are then to a greater or less extent available to the host. Herbivores as a class, therefore, have smaller amino-acid and vitamin requirements than pure carnivores. The total protein requirement of sheep can be reduced somewhat by the administration of urea, which the symbionts can utilise for the production of their own tissue protein. Similarly the milk yield of cows can be augmented by feeding urea, but how far the symbionts can contribute towards the total maintenance of the host has not been determined. In some animals, probably, the intestinal symbionts can produce all the amino-acids and other indispensable supplements the host requires and, according to some old and much-criticised experiments, this is true of cockroaches and black-beetles. Both these insects harbour intestinal symbionts and, according to the experimental results, they can be reared from the egg and brought to full sexual maturity through several generations with glycine, and perhaps even with ammonium sulphate, as the sole dietary source of nitrogen. This, undoubtedly, is due to the activities of the symbionts and it is to the same source that many insects presumably look for supplies of essential vitamins, for many such organisms live habitually on diets which are unlikely to contain all the necessary vitamins, even in minute traces. Probably something similar accounts for the survival of detritus-eaters like the earthworm, whose food is already so far decomposed as to contain little if any of the vitamins which even earthworms must presumably require. In some cases, too, the consumption and digestion of living bacteria and yeasts along with the food must play a significant part in the nutrition of animals.

Parasitism

As we have seen, organisms that have lived for many generations in environments containing important nutrient substances are liable to lose their power of synthesising these compounds. Organisms of this kind accordingly become restricted to certain kinds of environments and it is very probable indeed that highly parasitic bacteria, many of which are also highly pathogenic, are parasitic because they have gradually lost the ability to synthesise one after another of the numerous substances required for their own metabolism, and which they find ready-made in the body fluids and tissue-exudates which they normally inhabit. *Haemophilus influenzae*, *Streptococcus haemolyticus* and many other highly exacting pathogens are parasitic because their nutritional requirements are only likely to be fully met by the blood and tissue fluids of a living host. Having once become parasitic, a given organism is likely to lose the ability to synthesise more and more of the compounds it requires; in other words, the more parasitic it is, the more parasitic it is likely to become.

Most parasitic of all are the viruses which attack so many animals, plants and even micro-organisms. They have no metabolism of their own but, having gained entry into their hosts, for which they are highly specific, these viruses take over and completely reorganise and redirect the host's metabolism in such a way that the host no longer goes about its ordinary affairs. Instead it concentrates all its energies on *the production of more of the virus*!

Animals that parasitise plants are in an enviable position. If they are so morphologically specialised as to be able to get and maintain a tight hold upon their host, they have at their disposal all the vast range of compounds that plants can synthesise but animals cannot. We ourselves, in fact, are as 'parasitic' upon green plants as any plant louse; the only differences are, first, that we do our parasitising at a distance, and secondly, as Elton has pointed out, that while the parasite lives upon the host's income, we live upon its capital.

Organisms that parasitise animals are less secure, for their own existence presupposes the host's ability to secure enough of its own essential amino-acids and vitamins. We have seen enough to anticipate that these essential metabolites are likely to be precisely those upon which the parasite itself must largely depend. It is therefore to be anticipated that parasites that are too voracious or too numerous will be liable to bring about their own downfall by starving out their host. In many cases of heavy parasitic infestation we find indications of general, though on the whole rather indefinable, poor health and general condition. This has usually been used to argue for the production by the parasite of a toxin of some kind. While this is undoubtedly true of many parasitic bacteria, there is little convincing evidence that similar toxins are produced by animal parasites. It seems very probable, in fact, that the low level of general health would prove, if the condition were to be carefully investigated, to be due to a general nutritional deficiency caused by the diversion to its own purposes by the parasite of material for which the host, because of its own synthetic disabilities, is obliged to rely upon its food.

The biochemical study of parasitism has always been bedevilled by the difficulty, amounting usually to an impossibility, of cultivating parasites independently of their respective hosts. We know virtually nothing about the metabolism of most animal parasites and even less about their influence upon that of the host. In some cases the effects of infestation are very startling and highly specific, and this is true, for example, of crabs parasitised by *Sacculina*. This organism attacks crabs and its effect upon the males is one of feminisation; female crabs are little affected. Is it going too far to suspect that one of the nutritional supplements required by *Sacculina* can only be provided by whatever substance fulfils the role of a male sex hormone among crabs, or some intermediary that is normally involved in its elaboration?

The recent brilliant advances in methods for the culture of highly parasitic micro-organisms are of the utmost interest here. Starting with complicated mixtures it has usually been possible to find empirically some complex medium on which even the most exacting of microbes will live. By progressive simplification of this medium and systematic replacement of complex by simpler constituents it has been possible in many cases to get the organisms to grow on media that are still complex, but of which the composition is precisely known. Certain micro-organisms, e.g. *Haemophilus influenzae*, can survive on media containing blood, a highly complex mixture, but on removal of the whole blood and its replacement by one after another of many of its known constituents, it has been shown that the essential constituent is haematin. Similar procedures have revealed the identities of the growth factors required by a large variety of microbial organisms.

A similar mode of attack upon animal parasites might reasonably be expected to yield profitable results, but there still remains as a prime necessity the ability to culture the parasite apart from its host. True, it is possible to keep a large, robust form like the roundworm, *Ascaris*, in the laboratory for several days and to try the effects of different substances upon its survival. Many such experiments have in fact been done, but so far have led nowhere. But it is difficult to believe that the parasite remains in good physiological condition for more than a few hours after its removal from the intestine of the host, even though this worm maintains its motility for a week or more in individual cases. It is clear that we are not likely to learn very much about its metabolism until we can keep the parasite in isolation and, at the present time, this goal seems a long way off.

There are, however, encouraging signs. The eggs of another parasitic nematode, *Ostertagia*, can be reared on artificial media through the free-living larval stages, but as soon as the parasitic stage is reached no further maintenance has so far proved possible. With the development of the parasitic phase, some new factor enters into the problem; the metabolism changes from that of a free-living to that of a parasitic organism very suddenly, and if a way could be found of crossing the barrier in this one individual case the way might be thrown open to far-reaching advances in others.

The most promising results so far have been obtained in the case of another nematode, *Neoaplectana glaseri* which parasitises the Japanese beetle, *Popilia japonica*. Fully mature adult specimens of this nematode can be reared from the egg in artificial media through several generations. Eventually, however, the strain becomes infertile, and can only be restored by passing it through the normal host for several generations. The medium used in these experiments consisted of a veal-infusion agar jelly, together with glucose and a living culture of yeast, the latter being an admirable source of all the essential amino-acids and most of the known vitamins, especially the B group.

This important work is due to Glaser, and suggests that the principles that have already revealed so much about the nutrition of parasitic and highly exacting micro-organisms may yet prove to be applicable to other parasitic types. When at last we can break the at present binding link that ties the life of the parasite to that of its host, we may reasonably expect to learn something about the metabolism of the parasite and the nutritional bonds that confine it to this or that particular host. This, in its turn, will open the door to investigation of one of the most striking of all the problems of parasitism, namely the extraordinary degree of specificity of particular parasites towards particular hosts.

2. *Digestion*

Apart from sap-suckers, honey-eaters and the like, relatively few terrestrial animals take their food in solution. It was at one time widely believed, however, that a large part of the food of aquatic animals consists of dissolved matter, of which appreciable amounts are present in most natural waters. Pütter, the principal proponent of this view, carried out experiments over a period of many years which, he claimed, proved that natural waters do not contain enough particulate feeding stuffs to support the life of the animals he studied. Moreover, the quantity of dissolved organic matter far exceeds that in the particulate state. Pütter's work has, however, been sharply criticised, especially by Krogh, who points out that, among other things, Pütter underestimated the amount of particulate, and overestimated that of dissolved, material in natural waters, and his theory has now been generally discarded. Its place has been taken by the view that, in general, animals live almost wholly on particulate food. As Yonge has pointed out, if aquatic animals have to rely mainly upon dissolved substances, it is rather surprising that they should have developed the complex mechanisms for capturing, comminuting and digesting particulate food that they do in fact possess. Some of these mechanisms are exceedingly complicated and all of them are interesting. Certain animals make use of complex ciliary machinery so arranged as to pass streams of water over sheets of mucus, in which food particles become embedded to be swept by further ciliary movements into the mouth. Others, ranging from the great whale-bone whale to numerous microcrustacean species, make use of elaborate sieves which enable them to strain off particulate matter from the waters in which they live. Apart from devices such as these there are, of course, many specialised mechanisms for biting, boring, chewing, scraping or otherwise reducing large food masses into fragments of more manageable proportions.

The size of the food is an important factor in the organisation of the food relationships of animal communities. On the upper side the size is limited by the difficulty of capturing and killing prey that is relatively much larger than oneself, by the size and extensibility of one's oesophagus and so on, and on the lower side by the difficulty of catching enough organisms much smaller than oneself in a given time. Many ingenious devices have been evolved here too; some animals possess venomous secretions enabling them to overpower animals larger than themselves, while the whale-bone whale, which may reach a length of 80 feet or more, contrives to maintain its vast bulk on minute crustaceans, largely *Calanus*, collected by its 'strainers'. But the question of size is important in another connection, for in many cases the food must be disintegrated to a greater or less extent before it can be assimilated.

Speaking generally, assimilation may take place in either of two chief ways. It may occur by phagocytosis of particulate matter, followed by intracellular digestion, a method that calls for the pre-selection of food particles of a suitable size, and this is often accomplished by means of complicated ciliary sorting mechanisms. Alternatively, absorption may be preceded by extracellular digestion, which involves the secretion of extracellular enzymes by specialised digestive glands. These are in no way mutually exclusive processes; on the contrary, both mechanisms are often found to operate side by side in one and the same animal, and even in one and the same digestive organ. In any case it is doubtful whether there is any fundamental difference between the two mechanisms, for phagocytic absorption and the absorption of the dissolved products of extracellular digestion are perhaps only extreme forms of one and the same general phenomenon of permeability. The absorptive epithelia of the mammalian intestine, for example, are permeable only to small-molecular particles as far as amino-acids and sugars are concerned, yet they allow particles of undigested fat with diameters up to $0{\cdot}5\,\mu$ to pass through, provided that the particles are negatively charged.

The upper limit of size at which phagocytosis ceases to be possible does not seem to be sharply defined; probably, indeed, it varies considerably from one cell to another. But sooner or later, for any cell, a point must be reached at which phagocytosis is no longer possible, and assimilation must then depend upon previous comminution of what, in such cases, we may call the 'food mass'. The degree of dispersion necessary in a given case will depend upon the nature and properties of the cells or membranes through which absorption is to take place, and this dispersion may be effected by mechanical or by chemical means, or by both. Most usually both methods are employed. Extracellular digestion probably arose in response to the necessity of comminuting relatively large food masses as a preliminary to phagocytosis. Thus, apart from certain species which secrete extracellular digestive enzymes, no comminutive mechanisms are found in Protozoa or Porifera; feeding here depends upon preselection of food particles of the right size. The Coelenterata, however, are mainly carnivorous, and here the food is of larger size, being limited in dimensions only by the animal's ability to overpower organisms larger than itself and by the size of its rather extensible mouth. Mention has already been made of the paralysing 'sting' which many of these organisms are able to inflict (p. 77). The food mass is passed into the coelenteron (body cavity) and there attacked by a proteolytic enzyme which partially digests the proteinaceous components, breaks up the gross structure of the food mass and, with the assistance of contractions of the walls of the coelenteron, reduces the whole to a fine, particulate suspension. The dissolved products of digestion are absorbed by the cells of the coelenteron, which also take up the undigested particulate materials by phagocytosis, and digestion is then completed within the cells.

Phagocytic ingestion also plays an important part in digestion higher up in the animal kingdom. In the lamellibranch molluscs, for instance, extracellular digestion is confined to the disintegration of the starchy components of the food, the remainder being ingested by phagocytosis and intracellularly digested. In addition, an important part in assimilation is played by the wandering phagocytic cells that abound throughout the lamellibranch body: these seem to be capable of engulfing particles rather larger than can be dealt with by the cells of the digestive gland. But, with increasing complexity of structure and organisation, it seems that, on the whole, phagocytosis comes to play a progressively smaller and less important part in digestion. The absorptive powers of the digestive epithelia become restricted to smaller and smaller particles until, in the higher vertebrates, the food must be digested to yield very simple, water-soluble products before any absorption can take place. Among mammals, in fact, phagocytosis no longer plays any role in digestion and is confined to the wandering scavenger cells of the reticulo-endothelial system. The fats form an exception to this general rule however, for even in the mammals a large proportion of the total food fat is absorbed without previous hydrolysis. In the presence of bile salts, together with the products of partial digestion of a portion of the total fat, a large part of the whole is sufficiently finely dispersed to pass unchanged through the intestinal wall, and phagocytosis as such does not appear to be involved in the process. That this technique for fat absorption is not uniquely confined to the higher vertebrates is suggested by the discovery that highly surface-active substances, perhaps of the same chemical nature as the bile salts, are present in the gut contents of various Crustacea.

We know very little at the present time about the enzymes involved in intracellular digestion; indeed, the whole subject positively bristles with problems. In general a food particle, once ingested, becomes surrounded by a so-called digestive vacuole, the pH of which is usually on the acid side at first but later changes to neutrality or faint alkalinity. The initial acid phase has sometimes been regarded as favouring the destruction of bacteria which have been accidentally ingested along with the food, but it is not always present. Again, so powerful is the temptation to argue by analogy that the acid phase has been considered by some as corresponding to the gastric phase of digestion in the mammals, and the neutral or alkaline phase to pancreatic digestion. When digestion has been completed the vacuole disappears and the indigestible residues are expelled by a kind of reversed phagocytosis, often being clumped together before expulsion takes place.

Numerous digestively active enzyme preparations have been obtained by maceration of the phagocytic digestive glands of various animals, but it is always difficult to be certain that the enzymes extracted in this way are really enzymes of digestive function. Cells, probably of every kind, contain weak proteolytic and lipolytic enzymes, for example, and the proteolytic kathepsins of mammalian kidney or spleen tissue can digest proteins after extraction from the cells, but it is improbable in the highest degree that their normal intracellular function is a digestive one. However, strongly active preparations have been obtained on so many occasions that in some cases at least the enzymes can reasonably be suspected of being identical with those normally concerned with intracellular digestion. From the digestive gland of the oyster, *Ostrea*, for example, extracts have been prepared that give evidence of containing two proteolytic enzymes with pH optima at about 3·7 and 9·0 respectively. Even if we assume that these are enzymes of digestive function, it is still an open question as to whether or not they can possibly work under optimal conditions in the cell. On the one hand it can be urged that a pH of this order is not found even in the digestive vacuole itself but, on the other hand, the modern concept of the cell as a highly organised and essentially dynamic system lends support to the idea that it may well contain specialised regions in which a pH of a very unusual order could be achieved and maintained. But in any case, we cannot be sure that the observed optimum is necessarily the intracellular optimum. Destruction of the architectural features of the cell may involve contamination of the enzymes with interfering substances from which, in all probability, they are totally separated in the intact cell. Again, the enzymes are diluted in the course of extraction, and cases are known in which the optimum pH of an enzyme is altered by dilution, probably on account of the concomitant dilution of some indispensable activator or coenzyme. At present, therefore, it is impossible to go far into the question of intracellular digestion without skating on very thin ice; much of the work done is totally untrustworthy and there is little that can stand up to even the mildest criticism. It is probable in the extreme that, in order to obtain reliable evidence about the properties of the enzymes involved in intracellular digestion, it will be necessary to work with relatively pure preparations. This is a difficult undertaking, but not impossible of achievement. Already several mammalian kathepsins have been isolated, crystallised and studied in great detail and it is only a matter of time before the same can be done with intracellular enzymes of digestive function.

Considerably more is known about the enzymes involved in extracellular digestion, especially in the case of the mammals. We now have, for example, a wealth of information about the specificities and other properties of highly purified, crystalline preparations of pepsin, trypsin, chymotrypsin and a number of other digestive peptidases. It is unfortunate that the enzymes of animals other than the mammals have so far been almost untouched by modern techniques, but in the meantime one fact of the greatest interest has come to light. Comparative studies of extracellular mammalian peptidases and of the corresponding intracellular kathepsins have shown that the two groups are qualitatively and quantitatively homospecific, i.e. they have the same specificity requirements, in spite of the fact that in other respects the two groups show many important differences. So close are the resemblances of specificity, however, that it is difficult not to suspect that extracellular digestive enzymes had their origin in intracellular enzymes, and that extracellular digestion is, in effect, merely an exteriorised form of an intracellular process. That this is so seems all the more probable when we consider a third, rather rare type of digestive mechanism of which we have so far made no mention and which cannot be precisely classified either as intra- or as extracellular.

In some coelenterates and sometimes in higher phyla, digestion involves the formation of what are known as digestive syncytia. Among Platyhelminthes we find in the order Acoela that the gut is lined with highly amoeboid cells. When food is taken, these cells swell up and send out highly motile amoeboid processes that form a syncytial network in which the food mass becomes enmeshed. Digestion takes place within this network, which is later withdrawn. In some cases, however, it is withdrawn before digestion has proceeded very far, and digestion continues in its absence. In such cases it would appear that the essential function of the syncytium has been discharged once the secretory granules which it provides have been liberated. These secretory granules, according to histological evidence, are precursors of the digestive enzymes themselves. We have here a mechanism that shows every sign of being intermediate between the other two and encourages us to believe that extracellular digestion must have evolved from its intracellular counterpart by way of syncytial digestion. It is to be regretted that no chemical work has yet been done to follow up this important idea.

A great quantity of work has been published regarding the extracellular enzymes of animals in general, but little of it is of any great worth apart from the recent work on purified mammalian enzymes. The early work in particular is almost valueless to-day because, as J. B. S. Haldane has pointed out, work on enzymes prior to 1910 or thereabouts was 'at the mercy of unanalysable changes in the environmental conditions until Sörensen pointed out the dependence of enzyme activity upon pH'. Since Sörensen's observations, which were published in 1909, seem to have taken some years to penetrate into less exclusively biochemical parts of the biological literature, it is probably true to say that all the quantitative observations made before 1910 are in urgent need of repetition, and that those from 1910–14 would be the better for it. The war of 1914–18 put an effective end to research on problems of this kind, and it is only in work published since about 1920 that we can expect to find much reliable information except of a purely qualitative kind.

The bulk of the more recent work on digestive enzymes falls rather sharply into two categories. In the first we have work carried out by trained biochemists, but this has been mainly confined to mammalian enzymes and contains little of comparative interest. In the second category we find a vast mass of data, the bulk of which has been collected by comparative physiologists and zoologists whose interest and enthusiasm was, in a majority of cases, supported by little biochemical knowledge and less experience. Data of this kind fail in almost every case to furnish satisfactory information about the enzymes concerned, since the work has usually been done on the crudest of tissue extracts, and even then with little regard to even the most arbitrary standardisation of experimental conditions. In consequence, these results carry little conviction as evidence of the identity of any given enzyme and, at the same time, furnish little information regarding the probable efficiency of the enzyme in the normal animal.

In spite of these difficulties and limitations, however, a broad survey of the great mass of work that has been published leaves certain fairly definite impressions. First, it may be said that where digestion is partly intra- and partly extracellular, the extracellular enzymes found normally depend upon the nature of the food and show no regard for the dictates of taxonomy. In animals that are carnivorous the extracellular enzymes are those that attack proteins; in herbivores, on the other hand, extracellular amylases are found. Secondly, it seems highly probable that animals of every kind possess, either intracellularly or in extracellular digestive secretions, an armoury of catalysts collectively capable of breaking down proteins into amino-acids, poly- into monosaccharides, and fats into their component fatty acids. Thirdly, it seems probable that digestion follows essentially the same chemical pathways in animals of every kind, in spite of the manifold differences of manipulative technique found in different groups and species; and fourthly, little can be safely said about the specificities, kinetics, chemical nature or identity of any of these digestive enzymes excepting where, as in the case of the mammalian peptidases for example, the techniques and concepts of modern enzymology have been brought to bear upon them.

This is not the place to go into details, either of the properties and mode of operation of the mammalian enzymes which have been adequately described in nummerous monographs and text-books, nor is it possible here to present even a fraction of the older work on the distribution and properties of the digestive enzymes of sub-mammalian forms of life. If the reader feels, as justifiably he may, that the comparative biochemistry of digestive processes is in a sorry state at the present time, he may find comfort in the knowledge that techniques are nowadays available that should enable us in time to answer many of the problems so far unsolved. If, by chance, a few readers here and there should feel any urge to take an active part in solving these riddles, then the functions of this chapter will have been more than well accomplished.

CHAPTER IX

Metabolism and the environment

In the preceding chapters an attempt has been made to build up a picture of some of the physico-chemical relationships that exist between the internal and external environments of animals. Two factors in particular have been considered, namely, the availability of water and of salts. Granted adequate provisions of these commodities, together with efficient mechanisms for controlling their ingress to and egress from the body, it seems that animals can contrive to live almost anywhere, in the sea, in fresh water, on dry land and even in deserts. In spite of the wide range of habitats in which animals are found to-day, they all maintain in their blood a solution of salts, the relative proportions of which vary remarkably little from species to species. We see, here, it seems, evidence for the existence of a common, and probably fundamental, chemical ground-plan to which the bloods and tissue fluids of all animals must conform. They conform, not because they will, but because they must, for appreciable departures from the normal, relative balance of ions are not tolerated for long by any of the many kinds of cells and tissues that have been studied.

Superimposed on this apparently essential fixity of relative ionic composition we find secondary and specific modifications that reflect adaptations to new external conditions. These modifications like the rest must presumably originate in genetic mutations.

The *relative* ionic composition of the bloods of different animals is always essentially the same, whether they inhabit the sea, the fresh waters or the dry land, but the *total* ionic composition, as reflected in the osmotic pressure, varies over a wide range, being far lower in freshwater and terrestrial animals than it is in marine forms. Further secondary features of another kind are found in the addition to the primitive saline of blood pigments which facilitate the rapid transportation of oxygen to tissues that require it and, at the same time, assist in the removal of carbon dioxide from tissues that produce it. At still higher levels of organisation, in the vertebrates generally and especially in the mammals, we find a new device, the blood pigment now being packed away inside red corpuscles to form a mobile tissue of which the function is essentially one of respiration. The oxygen capacity of these corpusculated bloods is far higher than that attained in bloods which carry their respiratory pigments in solution, and the corpuscles play a very important part in the transport of carbon dioxide as well. In particular, these cells serve to provide a medium, separated from the internal medium proper, in which the formation and buffering of hydrogen ions can proceed without endangering the constancy of the pH of the true internal medium. But all these are secondary, specific modifications, all of which are of greater or less survival value to their possessors, and have been superimposed on the basic and indispensable ground-plan in the course of evolutionary time.

As a further example of the same kind we may consider the phenomena of digestion. In spite of the manifold differences that exist between the mechanisms for collecting and comminuting food, the chemical basis of the digestive processes seems always to be essentially the same. Complex food molecules are hydrolysed by teams of enzymes which, like proteins of other kinds, are probably species-specific, but appear nevertheless to be of much the same kind in all animals. There does in fact seem to be a general chemical ground-plan of digestion, and here again we sometimes find specialisation in particular cases, notably in the occurrence of cellulases in certain herbivorous molluscs.

We have had something to say also on the subject of nutrition from the comparative point of view. The striking fact that emerges here is that, so far as we have evidence, animals of every kind have substantially the same nutritional requirements; requirements, moreover, which they share with a great variety of micro-organisms. Thus of the B group of vitamins, there is evidence that several individual members are required not only by animals but by numerous bacteria as well. Furthermore, these substances are found in plant tissues and also in the tissues of other autotrophes which, like the plants, are capable of achieving their synthesis. Now there is abundant evidence that riboflavin, nicotinic amide, thiamine and pyridoxal form parts of certain essential enzymes and enzymic systems, and we are therefore drawn inevitably to the conclusion that the catalytic machinery, and therefore a large part of the metabolism, of plants, bacteria and animals alike, is essentially the same, again suggesting that there exists a common and fundamental metabolic ground-plan to which living organisms of every kind must probably conform. But here again, over and above the fundamental ground plan, we find secondary and specific adaptational features, for the vertebrates alone among animals are definitely known to require vitamin A which, as we have pointed out, forms a functional part of their visual apparatus. Plants, on the other hand, commonly contain and hence presumably employ a considerable number of carotenoid pigments for which animals, in general, have no immediate use, apart from personal adornment and as a source of vitamin A for those animals that require it.

Further evidence for the existence of a common metabolic ground-plan comes from consideration of the mechanisms involved in cellular respiration. Here we find dehydrogenases, coenzymes, flavoproteins, cytochrome oxidase and cytochrome itself all playing essential roles, and in cells and tissues of practically every kind that have been studied there is evidence that tissue respiration follows substantially the same lines. Occasionally we find that cytochrome is replaced by some other pigment of analogous function but, in the main, the mechanisms are always very much the same. As a secondary and specific feature we find that cytochrome is absent from strictly anaerobic bacteria, an indication that secondary modifications may sometimes be differences of a negative instead of a positive kind, of omission instead of addition. Another example of this 'negative specialisation', though its function in this case is still obscure, is found in the disappearance of the series of uricolytic enzymes as we pass up the evolutionary scale. Urico-oxidase, allantoinase, allantoicase and even urease are commonly present among invertebrates, but in the fishes allantoicase is often absent. By the time we reach the mammals allantoinase too has been lost, and finally, in man and the other Primates, even urico-oxidase has disappeared.

If now we turn to a consideration of the nitrogenous metabolism of animals we find yet more evidence. Proteins, after their digestive degradation to yield amino-acids, lose their amino-nitrogen by processes of de-amination. These processes may, for all we know, differ in detail in different animal groups but, in the end, they all yield the same end-product, ammonia. This appears to be universal among the animals. But here again we find adaptational modifications, this time in response to changes in the availability of water, some terrestrial animals converting their waste ammonia into urea and others again into uric acid.

Similar facts come to light in connection with the metabolism of carbohydrates, which seems always to centre round glucose and glycogen. The presence of glycogen or something very like it in behaviour and properties has been demonstrated in innumerable kinds of animal materials, though no systematic survey of the animal kingdom seems yet to have been attempted. Specimens of authentic glycogen have been isolated from a number of invertebrate sources, as well as from the livers and muscles of vertebrates and, although samples from different sources vary considerably in physical properties, there are no fundamental chemical differences between them. Apart from glycogen, few polysaccharides have very wide distribution. We have already alluded to the occurrence of cellulose in the body-wall of Tunicata, the one well-authenticated case of an animal cellulose. Other alleged cases are based mainly or entirely on evidence of histological staining, a notoriously unspecific and unreliable method. In the Tunicata the function of cellulose which, like glycogen, is a polyglucose, is essentially a structural and not a metabolic one. Another rare animal polysaccharide is the galactogen that occurs in the eggs and albumin glands of snails. This is of some chemical interest since it contains the rare sugar L-galactose in the proportion of 1 molecule for every 6 molecules of the ordinary D-form; here the function appears to be specifically concerned with reproduction. Finally, among animal polysaccharides we have chitin, a widely distributed substance derived from *N*-acetyl-glucosamine, of which the function again is essentially a structural one. Chitin forms a matrix which, impregnated with calcium carbonate and phosphate together with smaller amounts of other salts, furnishes the carapaces of the crabs and lobsters, or, when impregnated with the curious 'insect waxes', provides the impermeable, waterproof cuticle of the insects. Considering animal polysaccharides as a whole we are once more compelled to the conclusion that there exists a fundamental ground-plan that involves glycogen, and that occurrences of polysaccharides of other kinds are secondary features associated with specific, as opposed to fundamental tasks, cellulose and chitin being associated with structural, and galactogen with reproductive, phenomena.

Thus we arrive from several different lines of approach at essentially the same conclusion; that *there exists a common, fundamental chemical ground-plan of composition and metabolism to which all animals*, and very probably other living organisms also, *conform, and that, superimposed on these foundations, there are numerous secondary, specific and adaptational variations*, some of addition and others of omission. The further biochemistry goes the more does this appear to be true; indeed, many biochemists regard it as almost axiomatic.

This is not a hypothesis that is susceptible of direct proof or disproof in the present state of our knowledge but it does provide us with that most valuable of conceptual tools, a working hypothesis. As the late Sir F. G. Hopkins wrote in his foreword to the first edition of this book, one of the ultimate tasks of comparative biochemistry is 'to decide on what, from the chemical standpoint, is essential...as distinct from what is secondary and specific. For any such decisions the necessary harvest of contributory facts must come from many diverse fields.' The material covered by this little book provides a considerable collection of contributory facts, but not enough to permit us to arrive at any final decisions. Julian Huxley has shown that there appears to be a fundamental morphological ground-plan of animal growth and Joseph Needham that there appears to be a fundamental chemical ground-plan of animal growth and embryonic development. Are these merely two special cases of the general proposition propounded here, that all life is based on a fundamental pattern that is common to living organisms of every kind, and that all the immense diversity of shapes, sizes, habits, habitats and the rest is exclusively due to an agglomeration of secondary and specific, adaptational variations on the fundamental theme?

INDEX